CAMBRIDGE LIBRARY COLLECTION

Books of enduring scholarly value

Physical Sciences

From ancient times, humans have tried to understand the workings of the world around them. The roots of modern physical science go back to the very earliest mechanical devices such as levers and rollers, the mixing of paints and dyes, and the importance of the heavenly bodies in early religious observance and navigation. The physical sciences as we know them today began to emerge as independent academic subjects during the early modern period, in the work of Newton and other 'natural philosophers', and numerous sub-disciplines developed during the centuries that followed. This part of the Cambridge Library Collection is devoted to landmark publications in this area which will be of interest to historians of science concerned with individual scientists, particular discoveries, and advances in scientific method, or with the establishment and development of scientific institutions around the world.

The Life and Letters of Faraday

Michael Faraday (1791-1867) made foundational contributions in the fields of physics and chemistry, notably in relation to electricity. One of the greatest scientists of his day, Faraday held the position of Fullerian Professor of Chemistry at the Royal Institution of Great Britain for over thirty years. Not long after his death, his friend Henry Bence Jones attempted 'to join together his words, and to form them into a picture of his life which may be almost looked upon as an autobiography.' Jones' compilation of Faraday's manuscripts, letters, notebooks, and other writings resulted in this *Life and Letters* (1870) which remains an important resource for learning more about one of the most influential scientific experimentalists of the nineteenth century. Volume 1 (1791–1830) covers Faraday's earliest years as an errand boy and bookbinder's apprentice, his arrival at the Royal Institution as an assistant and his early publications on electricity.

Cambridge University Press has long been a pioneer in the reissuing of out-of-print titles from its own backlist, producing digital reprints of books that are still sought after by scholars and students but could not be reprinted economically using traditional technology. The Cambridge Library Collection extends this activity to a wider range of books which are still of importance to researchers and professionals, either for the source material they contain, or as landmarks in the history of their academic discipline.

Drawing from the world-renowned collections in the Cambridge University Library, and guided by the advice of experts in each subject area, Cambridge University Press is using state-of-the-art scanning machines in its own Printing House to capture the content of each book selected for inclusion. The files are processed to give a consistently clear, crisp image, and the books finished to the high quality standard for which the Press is recognised around the world. The latest print-on-demand technology ensures that the books will remain available indefinitely, and that orders for single or multiple copies can quickly be supplied.

The Cambridge Library Collection will bring back to life books of enduring scholarly value (including out-of-copyright works originally issued by other publishers) across a wide range of disciplines in the humanities and social sciences and in science and technology.

The Life and Letters of Faraday

VOLUME 1

BENCE JONES

CAMBRIDGE UNIVERSITY PRESS

Cambridge, New York, Melbourne, Madrid, Cape Town, Singapore,
São Paolo, Delhi, Dubai, Tokyo

Published in the United States of America by Cambridge University Press, New York

www.cambridge.org
Information on this title: www.cambridge.org/9781108014595

This edition first published 1870
This digitally printed version 2010

ISBN 978-1-108-01459-5 Paperback

Engraved by H. Adlard from a Photograph by Maull & Polyblank.

Most faithfully Yours
M. Faraday

London Longmans & Co.

THE

LIFE AND LETTERS

OF

FARADAY.

BY

DR. BENCE JONES,

SECRETARY OF THE ROYAL INSTITUTION.

IN TWO VOLUMES.

VOL. I.

SECOND EDITION, REVISED.

LONDON:

LONGMANS, GREEN, AND CO.

1870.

PREFACE

TO

THE SECOND EDITION.

In consequence of suggestions in letters and in reviews, some changes have been made in this edition.

Very little new matter has been added; but some letters have been left out; and other letters, and some of the lectures and journals have been shortened.

Two or three errors, which came from misapprehensions in conversation, have been corrected.

The most important mistake relates to the loaf of bread which Faraday had weekly when nine years old. I wrongly understood that it came from the temporary help which was given to the working class in London during the famine of 1801. I was too easily led into this error by my wish to show the height of the rise of Faraday by contrasting it with the lowliness of his starting point. I ought to have been content with the few words which he wrote. "My education was of the most ordinary description, consisting of little more than the rudiments of reading, writing, and arithmetic, at a common day school. My hours out of school were passed at home" (in the mews) "and in the streets."

This leaves no doubt that Faraday rose from that large class which lives by the hardest muscular labour, and can give but little for mental food; and yet by his own brain-work he became in his day the foremost of that small class which, by the mind alone, makes the glory of humanity.

H. B. J.

March 18th, 1870.

PREFACE

TO

THE FIRST EDITION.

To WRITE a life of FARADAY seemed to me at first a hopeless work. Although I had listened to him as a lecturer for thirty years and had been with him frequently for upwards of twenty years, and although for more than fifteen years he had known me as one of his most intimate friends, yet my knowledge of him made me feel that he was too good a man for me to estimate rightly, and that he was too great a philosopher for me to understand thoroughly. I thought that his biographer should if possible be one who was his own mental counterpart.

I afterwards hoped that the Journals, which he wrote at different periods whilst abroad, might have been published separately. If this had been done, then some portions of his biography would have been in his own writing: but it was thought undesirable to divide the records of the different parts of his life.

As time went on, and those who were most interested in the work found no one with sufficient leisure to whom they were inclined to give his manuscripts, I at last made the attempt to join together his own words, and to form them into a picture of his life which may almost be looked upon as an autobiography.

My first work was to read his manuscripts; and then to collect from his friends all the letters and notes that were likely to be of interest. And here, in duty bound, I must first thank Mrs. Faraday and her nieces Miss Barnard and Miss Reid for their help; then his earliest friend Mr. Abbott, whose collection of letters was priceless; then his friends M. Auguste de la Rive and the late Professor Schönbein. I am also indebted to Madame Matteucci, Miss Moore, Miss Magrath, Miss Phillips, Dr. Tyndall, Dr. Percy, Col. Yorke, the late Rev. John Barlow, and to many others.

From his letters, his laboratory note-books, his lecture-books, his Trinity House and other manuscripts, I have arranged the materials for a memorial of Faraday in the simplest order, with the least connecting matter.

I have, however, with permission, used some of the admirable summaries published by Dr. Tyndall, in his account of 'Faraday as a Discoverer.'

H. B. J.

October 18th, 1869.

CONTENTS

OF

THE FIRST VOLUME.

CHAPTER I.

1791–1812. To ÆT. 21.

PAGE

EARLY LIFE—ERRAND BOY AND BOOKBINDER'S APPRENTICE . . 1

CHAPTER II.

1812–1813. To ÆT. 22.

JOURNEYMAN BOOKBINDER AND CHEMICAL ASSISTANT AT THE ROYAL INSTITUTION 39

CHAPTER III.

1813–1815. To ÆT. 24.

EXTRACTS FROM HIS JOURNAL AND LETTERS WHILST ABROAD WITH SIR HUMPHRY DAVY 73

CHAPTER IV.

1815–1819. To ÆT. 28.

EARLIER SCIENTIFIC EDUCATION AT THE ROYAL INSTITUTION—FIRST LECTURES AT THE CITY PHILOSOPHICAL SOCIETY—FIRST PAPER IN THE 'QUARTERLY JOURNAL OF SCIENCE' 189

CHAPTER V.

1820–1830. To ÆT. 39.

HIGHER SCIENTIFIC EDUCATION AT THE INSTITUTION—MARRIAGE —FIRST PAPER IN THE 'PHILOSOPHICAL TRANSACTIONS' . . 276

LIST OF ILLUSTRATIONS.

FARADAY (JULY 27, 1857) WITH HIS HEAVY GLASS THAT SHOWED 'THE ACTION OF MAGNETISM ON LIGHT' . . *Frontispiece*

CLAPHAM WOOD HALL, YORKSHIRE, AS IT WAS . . . *page* 2

CLAPHAM WOOD HALL AS IT NOW IS 3

JACOB'S WELL MEWS, THE EARLY HOME OF FARADAY, AS IT NOW IS 7

THE BOOKBINDER'S SHOP IN BLANDFORD STREET AS IT WAS . . 9

LIFE OF FARADAY.

CHAPTER I.

EARLY LIFE—ERRAND BOY AND BOOKBINDER'S APPRENTICE.

THE village of Clapham, in Yorkshire, lies at the foot of Ingleborough, close to a station of the Leeds and Lancaster Railway. Here the parish register between 1708 and 1730 shows that 'Richard ffaraday' recorded the births of ten children. He is described as of Keasden, stonemason and tiler, a 'separatist;' and he died in 1741. No earlier record of Faraday's family can be found.

It seems not unlikely that the birth of an eleventh child, Robert, in 1724, was never registered. Whether this Robert was the son or nephew of Richard cannot be certainly known: however, it is certain that he married Elizabeth Dean, the owner of Clapham Wood Hall.

This Hall was of some beauty, and of a style said to be almost peculiar to the district between Lancaster, Kirkby Lonsdale, and Skipton. The porch had a gable-end and ornamented lintel with the initials of the builder (the proprietor); and the windows, with three or four mullions and label or string-course, had a very good effect. It was partly pulled down some twenty years ago, and a common sort of farm-house built in its place.

It is now little better than a stone cottage. The door opens directly into a kitchen, flagged with four large flags. What remains of the old Hall is, if

CLAPHAM WOOD HALL WITH MILL AS IT WAS.

anything, meaner than the dwelling itself. At this Hall Robert and Elizabeth Faraday lived, and had ten children, whose names and birthdays, and callings in after life, so far as they are known, were these :—

Robert; born 1724, died 1786, married 1756 to Elizabeth Dean of Clapham Wood Hall.

- Richard, born June 16, 1757, was an innholder, slater, grocer.
- John, born May 19, 1759, was a farmer.
- James, born May 8, 1761, was a blacksmith.
- Robert, born February 3, 1763, was a packer in a flax mill.
- Elizabeth, born June 27, 1765.
- William, born April 20, 1767, died in July 1791.
- Jane, born April 27, 1769.
- Hannah, born August 16, 1771.
- Thomas, born November 6, 1773, kept a shop.
- Barnabas, whose birthday is not known, was a shoemaker.

The first insight into this large family comes in the year when Faraday was born, through William, who died when he was twenty-four years old, at Clapham Wood Hall. Faraday's grandmother then wrote a letter to Anne Fordyce, to whom her son William

CLAPHAM WOOD HALL AS IT IS.

was engaged to be married. This letter shows the nature and strength of the religious feeling in the family for two generations previous to the birth of Faraday.

'Clapham Wood Hall, July 4, 1791.

'Dear Nancy,—With a troubled mind I write this to you. My dear son is dead. He died on the Sabbath in the evening at seven o'clock. Now, my dear love, I beg you would hear me what I have to say, and be sober. It hath been a great concern on William's mind about you: he was afraid you would feel to an extreme, and it troubled him very much: from this consideration he strove to make all things look as well as he could, and he had some hope within a little of his death that he happen might mend, which is very natural for all people.

'When William began to be worse, he began to be concerned about his everlasting welfare. He sent for Mr. Gorrel and confessed the faith in Christ, and gave Mr. Gorrel and the rest of the brethren great satisfaction.

'William was exceedingly comfortable, and rejoiced exceedingly. He then sent for his clothes, and he thought he would go to Wenning Bank, and join the brethren in public; but both we and the brethren saw there was no chance, but they came to visit him very frequently. I cannot, in a little compass, tell you all that William said, but he rejoiced exceedingly.

'Now, my dear love, I hope you will consider that Providence knows better than we, and I hope this account will serve in some measure to reconcile you, and I shall be very glad to hear from you.

'My children all give their kind love to you. From your affectionate, well-wishing

'ELIZABETH FARADAY.'

The brethren were members of a Sandemanian congregation. The Glasites are said once to have had a chapel at Clapham, with a burial ground attached to it. At present the chapel is converted into a barn, and the windows are walled up. The unconsecrated burial ground is thrown open to the fields, but one or two headstones still remain against the wall of the building.

Several of these congregations were formed in different parts of England by the writings and preaching of Robert Sandeman, the son-in-law of the Reverend John Glas, a Presbyterian clergyman in Scotland. Thus the Church in London was formed in 1760. In 1763 the congregation at Kirkby Stephen numbered between twenty and thirty persons. Sandeman ultimately went to America to make his views known, and he died there in 1771.

In 1728 Glas was deposed by the Presbyterian Church Courts, because he taught that the Church should be subject to no league nor covenant, but be governed only by the doctrines of Christ and His Apostles. He held that Christianity never was, nor could be, the established religion of any nation without becoming the reverse of what it was when first instituted ; that Christ did not come to establish any worldly power, but to give a hope of eternal life beyond the grave to His people whom He should choose of His own sovereign will; that the Bible, and that alone, with nothing added to it nor taken away from it by man, was the sole and sufficient guide for each individual, at all times and in all circumstances ; that faith in the divinity and work of Christ is the gift of God, and that the evidence of this faith is obedience to the commandments of Christ.

There are two points of practice in the Church which, in relationship to the Life of Faraday, must be mentioned. One of these is the admission into the Church, the other is the election of elders.

Members are received into the Church on the confession of sin, and the profession of faith in the death and resurrection of Jesus Christ. This profession must be made before the Church in public. The elders first, and afterwards the other members, ask such questions as they think are necessary to satisfy the Church. Prayer is then offered up, a blessing is invoked upon the person received, and he is heartily welcomed and loved for the sake of the truth he has professed.

There must be a plurality of elders (presbyters or bishops) in each Church, and two must be present at every act of discipline. When a vacancy occurs, the elders suggest for election to the congregation one of its members who appears to answer the description of an elder in the New Testament. The election is made by the whole Church unanimously. Earnestness of feeling and sincerity of conviction are the sole requisites for the office, which is entirely unpaid.

With regard to other members of the large family that were born at Clapham Wood Hall, it is known that Faraday's uncle John had a quarry among the hills, and erected a shielding for the use of the men, which in some maps is marked as Faraday House, and the gill which runs by it, in the map of the Ordnance Survey of Westmoreland, is called Faraday Gill. His uncle Thomas was the father of Thomas Armat Faraday, who is now a draper and grocer at Clapham. His father James, who was a blacksmith, was married in 1786 to Margaret Hastwell, a farmer's daughter of Mallestang, near Kirkby Stephen. To James and Margaret Faraday four children were born:—

James; born 1761, died 1810, married 1786 Margaret Hastwell, born 1764, died 1838.	Elizabeth, born 1787. Robert, born 1788. Michael, born 1791. Margaret, born 1802.

James soon after his marriage came to London, and

1791.

Sept. 22. Born.

lived at Newington, in Surrey, where his third child, Michael, was born on September 22, 1791. For a short time his home was in Gilbert Street; but about 1796 he moved to rooms over a coach-house in Jacob's Well Mews, Charles Street, Manchester Square: he

HOUSE IN JACOB'S WELL MEWS.

then worked as a journeyman at Boyd's in Welbeck Street. He joined the Sandemanian Church after he came to London. His wife, though one of the congregation, never became a member of the Church.

During the distress of 1801, when corn was above

9*l.* the quarter, Michael, who was nine years old, was given by his parents one loaf weekly, and it had to last him for that time.

In 1807 James wrote to his brother Thomas at Clapham—'I am sorry to say I have not had the pleasure of enjoying one day's health for a long time. Although I am very seldom off work for a whole day together, yet I am under the necessity (through pain) of being from work part of almost every day.' . . . And then, after speaking of some Church matters, he says—'But we, perhaps, ought to leave these matters to the overruling hand of Him who has a sovereign right to do what seemeth good to Him, both in the armies of heaven and amongst the inhabitants of the earth.'

On July 29, 1809, he wrote to the same brother—'I never expect to be clear of the pain completely with which I am afflicted, yet I am glad to say that I am somewhat better than I formerly was. . . .

'We are about to remove very shortly, so that you will be good enough to direct your next as follows—18 Weymouth Street, near Portland Place, London.'

There he died on October 30, 1810.

Faraday's mother died in Islington, in March 1838. 'She was very proud of her son; so much so, that Faraday asked his wife not to talk to his mother so much about him or his honours, saying she was quite proud enough of him, and it would not be good for her. Usually she called him "my Michael." She would do nothing whatever without his advice, and was quite contented and happy in being supported wholly by him in her declining years. She had not had any advantages of education, nor was she able to

1803.
ÆT. 11–12.

enter at all into her son's pursuits. She was particularly neat and nice in her household arrangements, and exerted herself to the utmost for her husband and children.'

The home of Michael Faraday was in Jacob's Well Mews from the time he was five years old until he went to Blandford Street. Very little is known of his life during these eight years. He himself has pointed out where he played at marbles in Spanish Place, and

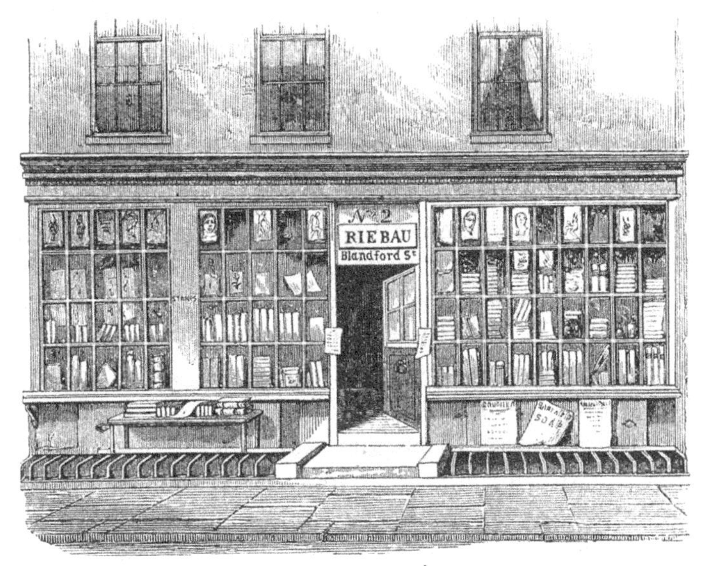

OLD VIEW OF RIEBAU'S SHOP.

where, at a later period, he took care of his little sister in Manchester Square. He says, 'My education was of the most ordinary description, consisting of little more than the rudiments of reading, writing, and arithmetic at a common day-school. My hours out of school were passed at home and in the streets.'

Only a few yards from Jacob's Well Mews is a bookseller's shop, at No. 2 Blandford Street.

There Faraday went as errand boy, on trial for a year, to Mr. George Riebau, in 1804. He has spoken with much feeling 'that it was his duty, when he first went, to carry round the papers that were lent out by his master. Often on a Sunday morning he got up very early and took them round, and then he had to call for them again; and frequently, when he was told the paper was not done with, "You must call again," he would beg to be allowed to have it; for his next place might be a mile off, and then he would have to return back over the ground again, losing much time, and being very unhappy if he was unable to get home to make himself neat, and to go with his parents to their place of worship.'

He says, 'I remember being charged with being a great questioner when young, but I do not know the nature of the questions.' One instance, however, has been preserved. Having called at a house, possibly to leave a newspaper, whilst waiting for the door to be opened, he put his head through the iron bars that made a separation from the adjoining house; and, whilst in this position, he questioned himself as to which side he was on. The door behind him being opened, he suddenly drew back, and, hitting himself so as to make his nose bleed, he forgot all about his question.

In after life the remembrance of his earliest occupation was often brought to his mind. One of his nieces says that he rarely saw a newspaper boy without making some kind remark about him. Another niece recalls his words on one occasion, 'I always feel a tenderness for those boys, because I once carried newspapers myself.'

Faraday's indentures as an apprentice are dated October 7, 1805: one line in them is worthy to be kept—'In consideration of his faithful service no premium is given.'

Four years later his father wrote (in 1809), 'Michael is bookbinder and stationer, and is very active at learning his business. He has been most part of four years of his time out of seven. He has a very good master and mistress, and likes his place well. He had a hard time for some while at first going; but, as the old saying goes, he has rather got the head above water, as there is two other boys under him.'

Faraday himself says, 'Whilst an apprentice I loved to read the scientific books which were under my hands, and, amongst them, delighted in Marcet's "Conversations in Chemistry," and the electrical treatises in the "Encyclopædia Britannica." I made such simple experiments in chemistry as could be defrayed in their expense by a few pence per week, and also constructed an electrical machine, first with a glass phial, and afterwards with a real cylinder, as well as other electrical apparatus of a corresponding kind.' He told a friend that Watts 'On the Mind' first made him think, and that his attention was turned to science by the article 'Electricity' in an encyclopædia he was employed to bind.

'My master,' he says, 'allowed me to go occasionally of an evening to hear the lectures delivered by Mr. Tatum on natural philosophy at his house, 53 Dorset Street, Fleet Street. I obtained a knowledge of these lectures by bills in the streets and shop-windows near his house. The hour was eight o'clock in the evening. The charge was one shilling per lecture, and my brother Robert (who was three years older and followed his father's business) made me a present of the money for several. I attended twelve or thirteen lectures between February 19, 1810, and September 26, 1811. It was at these lectures I first became acquainted with Magrath, Newton, Nicol, and others.'

He learned perspective of Mr. Masquerier,[1] that he might illustrate these lectures. 'Masquerier lent me Taylor's "Perspective," a 4to volume, which I studied closely, copied all the drawings, and made some other very simple ones, as of cubes or pyramids, or columns in perspective, as exercises of the rules. I was always very fond of copying vignettes and small things in ink; but I fear they were mere copies of the lines, and that I had little or no sense of the general effect and of the power of the lines in producing it.'

In his earliest note-book he wrote down the names of the books and subjects that interested him: this he called '"The Philosophical Miscellany," being a collection of notices, occurrences, events, &c., relating to the arts and sciences, collected from the public papers, reviews, magazines, and other miscellaneous works; intended,' he says, 'to promote both amusement and instruction, and also to corroborate or invalidate those theories which are continually starting

[1] Mr. Masquerier was probably a lodger in Mr. Riebau's house. In Crabb Robinson's Memoirs (vol. iii. p. 375, dated Feb. 18, 1851) it is written, 'At Masquerier's, Brighton. We had calls soon after breakfast. The one to be mentioned was that of Faraday. When he was young, poor, and altogether unknown, Masquerier was kind to him; and now that he is a great man he does not forget his old friend.'

into the world of science. Collected by M. Faraday, 1809–10.'

Among the books and subjects which are mentioned in this volume are, 'Description of a Pyropneumatic Apparatus,' and 'Experiments on the Ocular Spectra of Light and Colours,' by Dr. Darwin, from *Ackerman's Repository*; 'Lightning,' and 'Electric Fish and Electricity,' from *Gentleman's Magazine*; 'Meteorolites,' from the *Evangelical Magazine*; 'Water Spouts,' from the *Zoological Magazine*; 'Formation of Snow,' from *Sturm's Reflections*; 'To loosen Glass Stopples,' from the *Lady's Magazine*; 'To convert two Liquids into a Solid,' 'Oxygen Gas,' 'Hydrogen Gas,' 'Nitric and Carbonic Acid Gas,' 'Oxymuriate of Potash,' from *Conversations in Chemistry*.

'Galvanism:' 'Mr. Davy has announced to the Royal Society a great discovery in chemistry—the fixed alkalies have been decomposed by the galvanic battery,' from *Chemical Observer*; 'Galvanism and a Description of a Galvanometer,' from the *Literary Panorama*.

Through Mr. Tatum, Faraday made the acquaintance of Mr. Huxtable, who was then a medical student, and of Mr. Benjamin Abbott, who was a confidential clerk in the city, and belonged to the Society of Friends. Mr. Huxtable lent him the third edition of 'Thomson's Chemistry,' and 'Parkes's Chemistry:' this Faraday bound for his friend. The earliest note of Faraday's that is known to exist was written this year to Mr. Huxtable. It shows a little of the fun and much of the gentleness of his writing at this time:—

'Dear Sir,—Tit for tat, says the proverb; and it is

1812. ÆT. 20–21.

my earnest wish to make that proverb good in two instances. First, you favoured me with a note a short time since, and I hereby return the compliment; and, secondly, I shall call "tit" upon you next Sunday, and hope that you will come and tea "tat" with me the Sunday after. In short, the object of this note is to obtain your company, if agreeable to your convenience and health (which I hope is perfectly recovered long before this), the Sunday after next.

'This early application is made to prevent prior claims; and I propose to call upon you this day week to arrange what little circumstances may require it.

'In hope that your health is as well as ever, and that all other circumstances are agreeable, I subjoin myself, Sir, yours,

'M. FARADAY.'

The following are among the few notes which Faraday made of his own life:—

'During my apprenticeship I had the good fortune, through the kindness of Mr. Dance, who was a customer of my master's shop and also a member of the Royal Institution, to hear four of the last lectures of Sir H. Davy in that locality.[1] The dates of these lectures were February 29, March 14, April 8 and 10, 1812. Of these I made notes, and then wrote out the lectures in a fuller form, interspersing them with such drawings as I could make. The desire to be engaged in scientific occupation, even though of the lowest kind, induced me, whilst an apprentice, to write, in my ignorance of the world and simplicity of my mind, to Sir Joseph

[1] He always sat in the gallery over the clock.

Banks, then President of the Royal Society. Naturally enough, "no answer" was the reply left with the porter.'

On Sunday, July 12, 1812, three months before his apprenticeship ended, he began to write to Benjamin Abbott, who was a year and a half younger than his friend; but Abbott had been at good schools and was well educated, and hence Faraday regarded him as the possessor of a knowledge far beyond his own. Throughout all his correspondence this deference to his friend's superior school knowledge is always to be seen. These letters Mr. Abbott has most fortunately kept, thinking that at some future time they would be invaluable records of his friend's youth. They show his thoughts when he was 'giving up trade and taking to science,' during the period when the greatest change in the course of his life took place. The first eight were written between July 12 and October 1 in this year, whilst he was still an apprentice in Blandford Street.

They possess an interest almost beyond any other letters which Faraday afterwards wrote. It is difficult to believe that they were written by one who had been a newspaper boy and who was still a bookbinder's apprentice, not yet twenty-one years of age, and whose only education had been the rudiments of reading, writing, and arithmetic.

Had they been written by a highly educated gentleman, they would have been remarkable for the energy, correctness, and fluency of their style, and for the courtesy, kindness, candour, deference, and even humility, of the thoughts they contain.

1812.
ÆT. 20-21.

FARADAY'S FIRST LETTER TO BENJAMIN ABBOTT.

'Dear A——, Ceremony is useless in many cases, and sometimes impertinent; now between you and me it may not be the last, yet I conceive it is the first: therefore I have banished it at this time. But first let me wish you well, and then I will proceed on to the subject of this letter. Make my respects, too, if you please, to Mr. and Mrs. A., and also to your brother and sister.

'I was lately engaged in conversation with a gentleman who appeared to have a very extensive correspondence: for within the space of half an hour he drew observations from two letters that he had received not a fortnight before—one was from Sicily and the other from France. After a while I adverted to his correspondence, and observed that it must be very interesting and a source of great pleasure to himself. He immediately affirmed, with great enthusiasm, that it was one of the purest enjoyments of his life (observe, he, like you and your humble servant, is a bachelor). Much more passed on the subject, but I will not waste your time in recapitulating it. However, let me notice, before I cease from praising and recommending epistolary correspondence, that the great Dr. Isaac Watts (great in all the methods respecting the attainment of learning) recommends it as a very effectual method of improving the mind of the person who writes and the person who receives. Not to forget, too, another strong instance in favour of the practice, I will merely call to your mind the correspondence that passed between Lord Chesterfield and his son. In general, I do not approve of the moral tendency of Lord Chesterfield's letters, but I heartily agree with him respecting the utility of a written correspondence. It, like many other good things, can be made to suffer an abuse, but that is no effectual argument against its good effects.

1812.
ÆT. 20–1.

. . . . . .

'On looking back, I find, dear A., that I have filled two pages with very uninteresting matter, and was intending to go on with more, had I not suddenly been stopped by the lower edge of the paper. This circumstance (happily for you, for I should have put you to sleep else) has "called back my wand'ring thoughts;" and I will now give you what I at first intended this letter should be wholly composed of—philosophical information and ideas.

'I have lately made a few simple galvanic experiments, merely to illustrate to myself the first principles of the science. I was going to Knight's to obtain some nickel, and bethought me that they had malleable zinc. I inquired and bought some—have you seen any yet? The first portion I obtained was in the thinnest pieces possible—observe, in a flattened state. It was, they informed me, thin enough for the electric stick, or, as I before called it, De Luc's electric column. I obtained it for the purpose of forming discs, with which and copper to make a little battery. The first I completed contained the immense number of seven pairs of plates!!! and of the immense size of halfpence each!!!!!!

'I, Sir, I my own self, cut out seven discs of the size of halfpennies each! I, Sir, covered them with seven halfpence, and I interposed between, seven, or rather six, pieces of paper soaked in a solution of muriate of soda!!! But laugh no longer, dear A.; rather wonder at the effects this trivial power produced. It was sufficient to produce the decomposition of sulphate of magnesia—an effect which extremely surprised me; for I did not, could not, have any idea that the agent was competent to the purpose. A thought here struck me; I will tell you. I made the communication between the top and bottom of the pile and the solution with copper wire. Do you conceive that it was the copper that decomposed the earthy sulphate—that part, I mean, immersed in the solution? That a galvanic effect took place I am sure; for both wires became covered in a short time with bubbles of some gas, and a continued stream of very minute bubbles, appearing like small particles, ran through the solution from the negative wire. My proof that the sulphate was decomposed was, that in about two hours the clear solution became turbid: magnesia was suspended in it.

'Seeing the great effect of this small power, I procured from Knight some plate zinc, or sheet zinc I think they call it, about the thickness of pasteboard; from this I cut out discs, and also obtained some sheet copper, and procured discs of that metal. The discs were about 1¾ inch in diameter. These I piled up as a battery, interposing a solution of the muriate of soda by means of flannel discs of the same size. As yet I have only made one trial, and at that time had, I believe, about eighteen or twenty pairs of plates. With this power I have decomposed the sulphate of magnesia, the sulphate of copper, the acetate of lead, and I at first thought also water, but my conclusions in that respect were perhaps too hastily made.

'I inserted the wires into a portion of water that I took out of the cistern, and of course, in a short time, strong action commenced. A dense—I may really say dense—white cloud of matter descended from the positive wire, and bubbles rose rapidly and in quick succession from the negative wire; but after a time I perceived that the action slackened: the white cloud was scarcely perceptible at the wire, though by the former action the lower part of the solution was perfectly opaque and the bubbles nearly ceased. I thought that the action of the battery was exhausted; but in philosophy we do not admit suppositions; and therefore, to prove whether the battery was inert, or whether any principle in the water was exhausted, I substituted a fresh portion of water for that which had been galvanised. Then the action commenced again, and went on as at first. The white precipitate again appeared, and bubbles rose as before; but after a while it ceased, as in the first instance.

'I make no affirmative conclusion from these phenomena, but this I presume, that the water was not decomposed. Our water comes through iron pipes, and is retained in a leaden cistern. I have also ascertained that it holds a small portion of muriatic acid, and have no doubt that it contains carbonic acid. Now, do you think that any part of the lead or iron (the lead I should rather fancy) is held in solution by the muriatic or carbonic acid, and that the bubbles are formed by the precipitation of the metal, whilst the acid—what a blunder! I mean that the bubbles are formed by the escape of the acid and the precipitate—is the metallic oxide? Explain this circumstance to me—will you?—either by your pen or your tongue. Another phenomenon I observed was this: on separating the discs from each other, I found that some of the zinc discs had got a coating—a very superficial one in some parts—of metallic copper, and that some of the copper discs had a coating of oxide of zinc. In this case the metals must both have passed through the flannel disc holding the solution of muriate of soda, and they must have passed by each other. I think this circumstance well worth notice, for, remember, no effect takes place without a cause. The deposition, too, of the oxide of zinc in the flannel was curious, and will tend to illustrate the passage of the metals from one side to the other. I cannot describe it with any effect, you must see it; but think of these things, and let me, if you please, Sir, if you please, let me know your opinion.

'And now, dear Sir, to conclude in a manner requisite for the occasion. I heartily beg pardon for thus intruding on your time, your patience, and your good sense. I beseech you, if you will condescend so far, to return me an answer on this occasion, and pray let the refusal of your correspondence be as gentle as possible. Hoping, dear A., that the liberty I have taken will not injure me in your good opinion, I cannot conclude better than by wishing you all the happiness you can enjoy, the completion of all your good and honest wishes, and full health until I communicate with you again, and for ever after.

'I am, dear A., yours sincerely,

'M. FARADAY.'

'Monday morning, July 13.

'Dear A——, I am just now involved in a fit of vexation. I have an excellent prospect before me, and cannot take it up for want of ability: had I perhaps known as much of mechanics, mathematics, mensuration, and drawing, as I do perhaps of some other sciences—that is to say, had I happened to employ my mind with these instead of other sciences—I could have obtained a place—an easy place, too, and that in London—at 5, 6, 7, 800*l.* per annum. Alas! alas! Inability. I must ask your advice on the subject, and intend, if I can, to see you next Sunday. . . .

'I am, dear A., yours sincerely,

'M. FARADAY.

'One necessary branch of knowledge would be that of the steam-engine, and, indeed, anything where iron is concerned. Paper out, pen worn down, so good-day to you.'

The second letter to Abbott is dated July 20, 1812, Monday evening, ten o'clock.

To an honest man, close buttoned to the chin,
Broad cloth without, and a warm heart within.

'Here I am, Sir, on the third page of my paper, and have not yet begun to answer your very kind, free, friendly, instructive, amusing, and very welcome letter; but now I will turn to it and "say my say." For the first part I thank you; and here note that I shall keep you to the following words, "But will not fail to give them a thorough investigation." I like your logic well. Philosophical accounts, scientific inquiries, humble trials. Ha, ha, ha, hah! Don't you charge me with ceremony yet, or whilst your style runs thus.

'I am exceedingly obliged to you for the observation and quotation you have given me respecting Cupid and galvanism, and return my most grateful thanks to you for the remedy you have pointed out to me against the attacks of the little god—demon, by Le Sage's pardon. You, no doubt, are aware that this is not the first time that he has been conquered by philosophy and science. The last-named person informs us very minutely in what manner he was shut up in a glass bottle, and rendered incapable of doing mischief. Oh that I were as wise as that Sage, that I could shut little cupids in glass bottles! What exquisite presents they would be to the ladies! and how irresistible would the fair sex be to all who knew not how to oppose them thus unarmed, though I must confess they are not quite so absolute since the discovery of this anti-amorous remedy, galvanism. You will not have forgotten, too, when we set the nitrous oxide in opposition to him; and since galvanism now aids the gas, it is not possible for the little urchin to keep his ground. Farewell to him. I am now going to set my piles in action, in which state I shall leave them all night; and in the morning I will note down what phenomena I shall perceive. . . . . . . Alas! alas! the salt-box is empty, and as it is too late to procure a fresh quantity, I shall wish you all health and happiness, and wish you a good-night.'

'Tuesday morning, half-past six o'clock,
and a fine morning.

'Good-day to you, Sir. I now intend to proceed on with my letter from the point where I left off; not exactly though, for as yet I have no salt, and I do not like to substitute any other solution or any acid, because I suspect both the acid and the alkali bear a part in the transmission of the metals. I am exceedingly obliged to you for your ideas on this subject, and I think I need not say I received it with good-will. I never yet, dear A., received anything from you but what I met with that feeling; and for the rest of the sentence, had I thought that your mind was so narrow as to be chagrined at seeing a better solution of this phenomenon from another person, I certainly should never have commenced this correspondence with you.

'I was this morning called by a trifling circumstance to notice the peculiar motions of camphor on water: I should not have mentioned the simple circumstance but that I thought the effect was owing to electricity, and I supposed that if you were acquainted with the phenomenon, you would notice it. I conceive, too, that a science may be illustrated by those minute actions and effects, almost as much as by more evident and obvious phenomena. Facts are plentiful enough, but we know not how to class them; many are overlooked because they seem uninteresting: but remember that what led Newton to pursue and discover the law of gravity, and ultimately the laws by which worlds revolve, was—the fall of an apple.

'My knife is so bad that I cannot mend my pen with it; it is now covered with copper, having been employed to precipitate that metal from the muriatic acid. This is an excuse—accept it.

'Tuesday evening, eleven o'clock.

'I have just finished putting the battery, as you term it, in action, and shall now let it remain for the night, acting on a solution of the muriate of ammonia. This is the disposition made: fifteen plates of zinc and as many of copper are piled up with discs of flannel interposed; fifteen other plates of each metal are formed into a pile with pasteboard, both it and the flannel being soaked in a solution of common salt. These two piles are connected together, and their combined action employed as I before stated. The flash from it, when applied to the gums or eyes, is very vivid, and the action on the tongue, when in contact with the edges, will not allow it to remain there.

'With respect to your second solution of the passage of the metals, I have not time at present to think of it, nor have I room to say more than that I thank you for all on that subject; wait till I have heard of your experiments. Good-night.

'Wednesday morning, six o'clock.

'I can now only state facts, opinions you shall have next time. On looking at the pile this morning, I found that the muriate of ammonia had been decomposed, the alkali separated at the negative wire and escaped; this was evident last night by the cloud it formed with muriatic acid. The acid acted on the copper wire, and a muriate of copper was formed; this was again decomposed; and now I find the negative wire covered with a vegetation of copper, and the positive wire eaten away very considerably. The solution is of a fine blue colour, owing to the ammoniate of copper. On turning to the piles, I found the action of one considerable, the other was exhausted: the first contained the flannel discs, and they were yet very moist; the other had the paper discs, and they were quite dry: of course you know why the action ceased? On looking to the state of the plates particularly, I found but one in the pile containing flannel that was in the state I before noticed, that is, it being zinc and possessing a coating of copper. In the paper pile not a single zinc plate was affected that way; the copper plates in both piles were covered very considerably with the oxide of zinc. I am aware with you that zinc precipitates copper, and that the metals are oxided, before solution, in acid: but how does that effect their mōtion from one disc to another in contrary directions? I must trust to your experiments more than my own. I have no time, and the subject requires several.

'M. FARADAY.'

His third letter to his friend Abbott is dated August 11, 1812.

. . . . . .

'I thank you for your electrical experiment, but conceive the subject requires a very numerous series and of very various kind. I intend to repeat it, for I am not exactly satisfied of the division of the charge so as to produce more than one perforation. I should be glad if you would add to your description any conclusion which you by them are induced to make. They would tend to give me a fairer idea of the circumstances.

'I have to notice here a very singular circumstance—namely, a slight dissent of my ideas from you. It is this. You propose not to start one query until the other is resolved, or at least "discussed and experimented upon;" but this I shall hardly allow, for the following reasons. Ideas and thoughts often spring up in my mind, and are again irrevocably lost for want of noting at the time. I fancy it is the same with you, and would therefore wish to have any such objections or unsolved points exactly as they appear to you in their full force—that is, immediately after you have first thought of them; for to delay until the subject in hand is exhausted would be to lose all the intervening ideas. Understand, too, that I preserve your communications as a repository into which I can dip for a subject requiring explanation, and therefore the more you insert the more will it deserve that name; nevertheless, I do not mean to desert one subject for another directly it is started, but reserve it as an after subject of consideration.

. . . . . .

'Sir H. Davy's book is, I understand, already published, but I have not yet seen it, nor do I know the price or size. It is entitled "Elements of Chemical Philosophy."

. . . . . .

'Definitions, dear A., are valuable things; I like them very much, and will be glad, when you meet with clever ones, if you will transcribe them. I am exceedingly well pleased with Dr. Thomson's definition of Chemistry; he calls it the science of insensible motions: "Chemistry is that science which treats of those events or changes in natural bodies which consist of insensible motions," in contradistinction to mechanics, which treats of sensible motions.

'How do you define idleness?

'I forgot to insert a query when at the proper place, though I think an investigation of it would be of importance to the science of chemistry, and perhaps electricity. Several of the metals, when rubbed, emit a peculiar smell, and more particularly tin. Now, smells are generally supposed to be caused by particles of the body that are given off. If so, then it introduces to our notice a very volatile property of those metals. But I suspect their electric states are concerned; and then we have an operation of that fluid that has seldom been noticed, and yet requires accounting for before the science can be completed.

'Health, happiness, and prosperity be with you; and believe me continually yours very sincerely,

'M. FARADAY.'

His fourth letter to Abbott is dated August 19, 1812.

'Dear A——, ... This letter will be a dull one, for I have but few subjects, and the heat of the weather has so enervated me that I am not able to treat those I have in a proper manner. But rouse up, Michael, and do not disgrace thyself in the opinion of thy friend.

'I have again gone over your letter, but am so blinded that I cannot see any subject except chlorine to write on; but before entering on what I intend shall fill up the letter, I will ask your pardon for having maintained an opinion against one who was so ready to give his own up. I suspect from that circumstance I am wrong . . . . With respect to chlorine, if we intend to debate the question of its simple or compound nature, we have begun at a wrong point, or rather at no point at all. Conscious of this, I will at this time answer your present objections but briefly, and then give the best statement I can of the subject. The muriate of soda is a compound of chlorine and sodium, and as chlorine in the theory is esteemed a simple substance, I conceive that the name of chlor*ate* of sodium is improper; *ate* and *ite* are the terminations of the generic name of salts, and convey to our minds an idea of the acid that the base is combined with. But chlorine is not an acid; it is a simple substance belonging to the same class as oxygen, and therefore its binary compounds should, I conceive, be termed, in imitation of oxides, chlorides. The muriate of soda is, therefore, a chlor*ide* of sodium, and the oxymuriate of soda is a compound of that chloride with oxygen.

'I will not say more at present on your objections, since you will now be able to answer them yourself in the same way that I should do; but I will proceed to the more simple and elementary parts of the subject. In the present case I conceive that experiments may be divided into three classes; 1st, those which are for the old theory of oxymuriatic acid, and consequently oppose the new one; 2nd, those which are for the new one, and oppose the old theory; and 3rd, those which can be explained by both theories—apparently so only, for in reality a false theory can never explain a fact. I am not aware of any belonging to the first class; what appeared to be such at first have on consideration resolved themselves into the third class; of the second class I will propose a few to you; and of the third class is that we have already been engaged upon.

'Be not surprised, my dear A., at the ardour with which I have embraced the new theory. I have seen Davy himself support it. I have seen him exhibit experiments, conclusive experiments, explanatory of it, and I have heard him apply those experiments to the theory, and explain and enforce them in (to me) an irresistible manner. Conviction, sir, struck me, and I was forced to believe him; and with that belief came admiration (then follow experiments).

. . . .

'I have not time, dear B., at present to close my

letter in a proper manner. I shall be at Ranelagh to-morrow evening (if fate permits); and if we do not meet before, will take my station exactly at nine under the orchestra. Yours truly,

'M. FARADAY.'

His fifth letter to Abbott was written September 9, 1812.

'You wrong me, dear A., if you suppose I think you obstinate for not coinciding in my opinion immediately; on the contrary, I conceive it to be but proper retention. I should be sorry indeed were you to give up your opinion without being convinced of error in it, and should consider it as a mark of fickleness in you that I did not expect. It is not for me to affirm that I am right and you wrong; speaking impartially, I can as well say that I am wrong and you right, or that we both are wrong, and a third right. I am not so self-opinionated as to suppose that my judgment and perception in this or other matters is better or clearer than that of other persons; nor do I mean to affirm that this is the true theory in reality, but only that my judgment conceives it to be so. Judgments sometimes oppose each other, as in this case; and as there cannot be two opposing facts in nature, so there cannot be two opposing truths in the intellectual world; consequently, when judgments oppose one must be wrong—one must be false, and mine may be so for aught I can tell. I am not of a superior nature to estimate exactly the strength and correctness of my own and other men's understanding, and will assure you, dear A., that I am far from being convinced that my own is always right. I have given you the theory—not as the true one, but as the one which appeared true to me—and when I perceive errors in it, I will immediately renounce it in part or wholly as my judgment may direct. From this, dear friend, you will see that I am very open to conviction; but from the manner in which I shall answer your letter, you will also perceive that I must be convinced before I renounce.

'You have made a blunder in your letter, A. You say that you will first answer my experiments, and then relate others; but you have only noticed one of mine, and therefore I suppose the answers to the others are to come. "With respect to the taper," do you mean to say that none of its carbon is burnt in atmospheric air or oxygen gas? I understood Davy that none of it was burnt in chlorine gas; and as for your query of water being formed, I do not believe there was any—not the slightest condensation took place. I did not insist much on this experiment by itself, but had connected it to another where charcoal would not burn. You should have answered them both together.

'Wednesday night, 10.30 P.M.

. . . . . .

'You wish to alter the tenor of our arguments; you conceive that if you prove oxygen to exist in muriatic acid you have done enough. Not so; if you do that you will do wonders, and I shall certainly pay that respect to it it deserves; but the experiments I have related must also be answered before I change opinions I understand. It is possible to support a new theory of chlorine—namely, that it is a compound of an unknown base and oxygen, but which has never yet been detected separate; but this will not alter our arguments, since still muriatic acid is considered as the chlorine and hydrogen united, and whilst this chlorine is undecomposed we must consider it as simple. I was considerably surprised to hear you last night charge me with having denied facts. I am not aware that I have denied any; nor do I wish to do so. I have denied some which have been accounted facts, but those cannot be what you alluded to. Pray point them out to me.

'I shall now answer all your conclusive experiments, and must confess I do not see that difficulty I expected. Do you remember the first experiment you quoted, the solution of a metal in muriatic acid, in which experiment you consider the metal as being oxidised at the expense of the acid? By this means you have arrived at a discovery which has drawn the attention of all great chemists—the decomposition of the muriatic acid; for by informing us what remains by the deoxidation of the acids by the metal, we shall have its other constituent part; and thus our dispute—no, not dispute, friendly controversy—will end.

'I fear, dear A., you will find it hard to decompose muriatic acid by the solution of a metal in it. It has never knowingly been done by any of Lavoisier's disciples yet; or, at least, they have never allowed it. It has been done, and I have before related the experiment to you. But, to return to your experiment. When a metal is dissolved in muriatic acid, I believe it is generally the case that hydrogen is evolved. From whence is the hydrogen but from decomposed water? and in what manner is the oxygen employed but by combining with the metal?—the oxide is then dissolved. As very prominent instances of this kind I will notice

1812. Æt. 20–1.

the action of muriatic acid on iron and zinc. Other metals are dissolved by this acid, but I have never noticed the phenomena attendant. If you say the metal obtains oxygen from the acid, inform me what part of the acid is left, and in what state.

'Secondly, oxygen, I know, may be obtained from the oxymuriates, because they contain it. They are formed by double combination: first a muriate is formed, being a compound of chlorine, and the metallic base of the alkalies, and with this compound oxygen combines. By applying heat, the only operation that takes place is the driving off of oxygen—but more of this when I have detailed further to you Davy's theory, though you must perceive the experiments are as easily explained thus as by Lavoisier's opinion.

'Thirdly, you can refer, I presume, to J. Davy's experiment, and therefore I shall give here only my opinion on it, whether mechanical or chemical. If the oxide is held mechanically in the ferrane, as he supposes, it makes no part of the compound of chlorine and iron, and, of course, does not affect the subject at all in my idea; and if chemically, which is not at all probable, it does not make its appearance until water is added, and then it is easily accounted for: but in order to estimate the experiments exactly, it will be necessary to consider the manner in which ferrane is formed.

'I come next to your remarks, of which I own the propriety; and though I do not suppose that at any time I can make experiments with more exactness and precision than those I have quoted, yet certainly the performance would give us a clearer idea. I accept of your offer to fight it out with joy, and shall in the battle experience and cause, not pain, but I hope pleasure; nevertheless, I will, if you will allow me, give, whilst I have time and opportunity, and whilst my ideas are fresh and collected, what little more I know of this theory; not requiring your immediate answer to it, but leaving it to your leisure consideration.'

(He then gives the views of Davy on chlorine at length.)

'It is now time to conclude, dear A., which I do with best wishes to yourself and friends. In my next I will conclude the subject with euchlorine, when I will again subscribe myself, your sincere friend,

'M. FARADAY.'

Postscript.

'Dear A——, I have received yours of to-day, the perusal of which has raised in my mind a tumult of petty passions, amongst which are predominant vexation, sorrow, and regret. I write under the influence of them, and shall inform you candidly of my feelings at this moment. You will see by the foregoing part of this letter that I have not acted in unison with your request by dropping the subject of chlorine, and for not having done so I feel very considerable sorrow. I had at various short intervals, as time would permit, drawn it up, and felt, I will own, gratified on reading it over; but the reception of yours has made me most heartily regret it. Pity me, dear A., in that I have not sufficiently the mastery of my feelings and passions. In the first part of this long epistle you will see the reasons I have given for continuing the subject, but I fancy that I can now see the pride and self-complacency that led me on; and I am fearful that I was influenced by thinking that I had a superior knowledge in this particular subject. Being now aware of this passion, I have made a candid confession of it to you, in hopes to lessen it by mortifying it and humiliating it. You will of course understand that I shall not now enter on euchlorine until it is convenient for both of us, when I hope to take up the subject uninfluenced by any of those humiliating, and to a philosopher disgraceful, feelings.

'I subscribe myself, with humility, yours sincerely,

'M. FARADAY.'

The sixth letter was written to Abbott, September 20, 1812.

'What? affirm you have little to say, and yet a philosopher! What a contradiction! what a paradox! 'tis a circumstance I till now had no idea of, nor shall I at any time allow you to advance it as a plea for not writing. A philosopher cannot fail to abound in subjects, and a philosopher can scarcely fail to have a plentiful flow of words, ideas, opinions, &c. &c., when engaged on them; at least, I never had reason to suppose you deficient there. Query by Abbott: "Then pray, Mike, why have you not answered my last before now, since subjects are so plentiful?" 'Tis neither more nor less, dear A., than a want of time. Time, Sir, is all I require, and for time will I cry out most heartily. Oh that I could purchase at a cheap rate some of our modern gents' spare hours, nay, days; I think it would be a good bargain both for them and me. As for subjects, there is no want of them. I could converse with you, I will not say for ever, but for any finite length of time. Philosophy would furnish us with matter; and even now, though I have said *nothing*, yet the best part of a page is covered.

'How prone is man to evil! and how strong a proof have I of that propensity when even the liberal breast of my friend A. could harbour the vice of covetousness! Nevertheless, on a due consideration of the cause, and a slight glance at my own feelings on the same subject (they will not bear a strict scrutiny), I pass it over thus——

. . . . . .

'Your commendations of the MS. lectures[1] compel me to apologise most humbly for the numerous—very very numerous—errors they contain. If I take you right, the negative words "no flattery" may be substituted by the affirmative "irony:" be it so, I bow to the superior scholastic erudition of Sir Ben. There are in them errors that will not bear to be jested with, since they concern not my own performance so much as the performance of Sir H. There are, I am conscious, errors in theory, and those errors I would wish you to point out to me before you attribute them to Davy.' . . . . .

His seventh letter to his friend was written September 28, 1812.

'Dear A——, . . . I will hurry on to philosophy, where I am a little more sure of my ground. Your card was to me a very interesting and pleasing object.[2] I was highly gratified in observing so plainly delineated the course of the electric fluid or fluids (I do not know which). It appears to me that by making use of a card thus prepared, you have hit upon a happy illustrating

[1] The notes of the lectures of Davy taken in the spring of this year.

[2] Many will remember the use he made of this experiment in after years in his lectures.

medium between a conductor and a non-conductor; had the interposed medium been a conductor, the electricity would have passed in connection through it—it would not have been divided; had the medium been a non-conductor, it would have passed in connection, and undivided, as a spark over it, but by this varying and disjoined conductor it has been divided most effectually. Should you pursue this point at any time still further, it will be necessary to ascertain by what particular power or effort the spark is divided, whether by its affinity to the conductor or by its own repulsion; or if, as I have no doubt is the case, by the joint action of these two forces, it would be well to observe and ascertain the proportion of each in the effect. There are problems the solution of which will be difficult to obtain, but the science of electricity will not be complete without them; and a philosopher will aim at perfection, though he may not hit it—difficulties will not retard him, but only cause a proportionate exertion of his mental faculties.

'I had a very pleasing view of the planet Saturn last week through a refractor with a power of ninety. I saw his ring very distinctly; 'tis a singular appendage to a planet, to a revolving globe, and I should think caused some peculiar phenomena to the planet within it. I allude to their mutual action with respect to meteorology and perhaps electricity.' . . . .

His eighth letter to Abbott is dated October 1, 1812; it was the last that he wrote before his apprenticeship ended.

'No—no—no—no, none; right—no, philosophy is not dead yet—no—O no; he knows it—thank you—'tis impossible—bravo!

'In the above lines, dear A., you have full and explicit answers to the first page of yours dated September 28. I was paper-hanging at the time I received it; but what a change of thought it occasioned; what a concussion, confusion, conglomeration; what a revolution of ideas it produced—oh! 'twas too much;—away went cloths, shears, paper, paste and brush, all—all was too little, all was too light to keep my thoughts from soaring high, connected close with thine.

'With what rapture would a votary of the Muses grasp that inimitable page! how would he dwell on every line and pore on every letter! and with what horror, dread, disgust, and every repulsive passion, would he start back from the word BARILLA to which I now come! I cannot here refrain from regretting my inability (principally for want of time) to perform the experiments you relate to me. I mean not to reflect on any want of clearness in your details; on the contrary, I congratulate you on the quickness with which you note and observe any new appearances; but the sight possesses such a superiority over the other senses, in its power of conveying to the mind fair ideas, that I wish in every case to use it. I am much gratified with your account of the barilla; but do I read right that part of your letter which says that the salt you obtained from the first treatment of it was *efflorescent*? As I went on to that passage, I did not expect that you would obtain any crystals at all, but only an uniform mass; but that crystals containing so great a quantity of alkali, in I suppose nearly a free state, should give out water to the atmosphere, surprised me exceedingly —explain, if you please.

. . . .

1812. ÆT. 21.

'I rejoice in your determination to pursue the subject of electricity, and have no doubt that I shall have some very interesting letters on the subject. I shall certainly wish to (and will if possible) be present at the performance of the experiments; but you know I shall shortly enter on the life of a journeyman, and then I suppose time will be more scarce than it is even now.

'Venus, I find, is amongst your visible planets—'tis a—beautiful—object—certainly.'

CHAPTER II.

JOURNEYMAN BOOKBINDER AND CHEMICAL ASSISTANT AT THE ROYAL INSTITUTION.

1812. ÆT. 21.

ON October 8, 1812, Faraday went as a journeyman bookbinder to a Mr. De la Roche, then a French emigrant in London. He was a very passionate man, 'and gave his assistant so much trouble that he felt he could not remain in his place,' although his master held out every inducement to him to stay, and even said to him, 'I have no child, and if you will stay with me you shall have all I have when I am gone.'

The letters which Faraday wrote to his friends Abbott and Huxtable show 'how eager was the desire he felt to proceed further in the way of philosophy,' and how strongly he was drawn towards 'the service of science;' they also show how far he had educated himself when he first went to the Royal Institution, and they give an insight into his character when he changed his course of life, and began his scientific career at the very lowest step, but under the greatest master of the time.

Four days after his apprenticeship ended, he wrote to his friend Abbott.

1812.
ÆT. 21.

'Sunday afternoon, October 11, 1812.

'Dear A——, I thank you heartily for your letter yesterday, the which gave me greater pleasure than any one I had before received from you. I know not whether you will be pleased by such commendation or not; it is the best I can bestow. I intend at this time to answer it, but would wish you, before you read the ensuing matter, to banish from your mind all frivolous passions. It is possible that what I may say would only tend to give rise (under their influence) to disdain, contempt, &c., for at present I am in as serious a mood as you can be, and would not scruple to speak a truth to any human being, whatever repugnance it might give rise to. Being in this state of mind, I should have refrained from writing to you, did I not conceive, from the general tenor of your letter, that your mind is, at proper times, occupied on serious subjects to the exclusion of those which comparatively are frivolous.

'I cannot fail to feel gratified, my dear friend, at the post I appear to occupy in your mind, and I will very openly affirm that I attach much greater importance to that interest since the perusal of your last. I would much rather engage the good opinion of one moral philosopher who acts up to his precepts, than the attentions and commonplace friendship of fifty natural philosophers. This being my mind, I cannot fail to think more honourably of my friend since the confirm ation of my good opinion, and I now feel somewhat satisfied that I have judged him rightly.

'As for the change you suppose to have taken place with respect to my situation and affairs, I have to thank my late master that it is but little. Of liberty and time I have, if possible, less than before, though I hope my circumspection has not at the same time decreased; I am well aware of the irreparable evils that an abuse of those blessings will give rise to. These were pointed out to me by common sense, nor do I see how any one who considers his own station, and his own free occupations, pleasures, actions, &c., can unwittingly engage himself in them. I thank that Cause to whom thanks are due that I am not in general a profuse waster of those blessings which are bestowed on me as a human being—I mean health, sensation, time, and temporal resources. Understand me clearly here, for I wish much not to be mistaken. I am well aware of my own nature, it is evil, and I feel its influence strongly; I know too that—but I find that I am passing insensibly to a point of divinity, and as those matters are not to be treated lightly, I will refrain from pursuing it. All I meant to say on that point was that I keep regular hours, enter not intentionally into pleasures productive of evil, reverence those who require reverence from me, and act up to what the world calls good. I appear moral and hope that I am so, though at the same time I consider morality only as a lamentably deficient state.

'I know not whether you are aware of it by any means, but my mind delights to occupy itself on serious subjects, and I am never better pleased than when I am in conversation with a companion of my own turn of mind. I have to regret that the expiration of my apprenticeship hath deprived me of the frequent company and conversation of a very serious and improving young companion, but I am now in hopes of a compensation by the acquisition of, at times, a letter from you. I am very considerably indebted to him for the sober turn or bent of my reason, and heartily thank him for it. In our various conversations we have frequently touched on the different parts of your letter, and I have every reason to suppose that, by so doing, we have been reciprocally benefitted.

'I cannot help but be pleased with the earnest manner in which you enforce the necessity of precaution in respect of new acquaintances. I have long been conscious of it, and it is that consciousness which limits my friends to the very small number that comprises them. I feel no hesitation in saying that I scrutinised you long and closely before I satisfied the doubts in my breast, but I now trust they are all allayed.

'It appears that in the article of experience you are my superior. You have been tried; if the result of the trial satisfies your own good sense and inward admonitions, I rest satisfied that you acted rightly. I am well aware that to act rightly is at times difficult; our judgment and good sense are oftentimes opposed, and that strongly too, by our passions and wishes. That we may never give up the first for the sake of the last is the earnest wish of your friend.

'I have made use of the term friend several times, and in one place I find the expression commonplace friendship. It will perhaps not be improper at this time to give you my ideas on true friendship and eligible companions. In every action of our lives I conceive that reference ought to be had to a Superior Being, and in nothing ought we to oppose or act contrary to His precepts. These ideas make me extremely displeased with the general and also the ancient idea of friendship. A few lines strike upon my mind at this moment; they begin thus :—

> A generous friendship no cold medium knows,
> But with one love, with one resentment glows, &c.

and convey sentiments that in my mind give rise to extreme disgust. According to what I have said, a few lines above, I would define a friend, a true friend, to be "one who will serve his companion next to his God;" nor will I admit that an immoral person can fill completely the character of a true friend. Having this idea of friendship, it was natural for me to make a self-inquiry, whether I could fill the character, but I am not satisfied with my own conclusions on that point; I fear I cannot. True friendship I consider as one of the sublimest feelings that the human mind is capable of, and requires a mind of almost infinite strength, and at the same time of complete self-knowledge. Such being the case, and knowing my own deficiency in those points, I must admire it, but fear I cannot attain it. The above is my opinion of true friendship, a passion or feeling I have never personally met with, and a subject that has been understood by very few that I have discussed it with. Amongst my companions I am conscious of only one who thinks the same of it that I do, but who confesses his inability to fill the character.

'When meditating and examining the character of a person with respect to his fitness for a companion, I go much farther than is generally the case. A good companion, in the common acceptation of the word, is one who is respectable both in connection and manners, is not in a lower rank of life than oneself, and does not openly or in general act improperly; this I say is the common meaning of the word, but I am by no means satisfied with it. I have met a good companion in the lowest path of life, and I have found such as I despised in a rank far superior to mine. A companion cannot be a good one unless he is morally so; and however engaging may be his general habits, and whatever peculiar circumstances may be connected with him, so as to make him desirable, reason and common sense point him out as an improper companion or acquaintance unless his nobler faculties, his intellectual powers, are, in proportion, as correct as his outward behaviour. What am I to think of that person who, despising the improvement and rectitude of his mind, spends all his efforts in arranging into a nice form his body, speech, habits, &c.? Is he an estimable character? Is he a commendable companion? No, surely not. Nor will such ever gain my commendation. On recollecting myself, I fancy I have said enough on this subject; I will therefore draw towards a conclusion.

. . . . . .

'I am in hopes of again hearing from you at some of your serious moments, at which time you, of course, will express yourself as I have done, without ceremony. But I must conclude in confidence that you are an eligible companion; and wishing that you may attain even to the character of a true friend,

'I remain yours, dear A., very sincerely,

'M. FARADAY.'

A few days later he writes to his friend Mr. Huxtable.

'London, October 18, 1812.

'Dear Huxtable,—You will be at a loss to know what to think of me, inasmuch as near two months have expired, and you have not, in that time, received any answer to your agreeable communication. I have to beg your pardon for such delay, and scarce know how satisfactorily to account for it. I have indeed acted unadvisedly on that point, for, conceiving that it would be better to delay my answer until my time was expired, I did so. That took place on October 7, and since then I have had by far less time and liberty than before. With respect to a certain place I was disappointed, and am now working at my old trade, the which I wish to leave at the first convenient opportunity. I hope (though fear not) that you will be satisfied with this cause for my silence; and if it appears insufficient to you, I must trust to your goodness. With respect to the progress of the sciences I know but little, and am now likely to know still less; indeed, as long as I stop in my present situation (and I see no chance of getting out of it just yet), I must resign philosophy entirely to those who are more fortunate in the possession of time and means.

'Sir H. Davy is at present, I believe, in Scotland. I do not know that he has made any further advances in Chemical science. He is engaged in publishing a new work, called "The Elements of Chemical Philosophy," which will contain, I believe, all his discoveries, and will likewise be a detail of his philosophical opinions. One part of the first volume is published. It is in price 11*s.* or 12*s.* 6*d.* I have not yet seen it. Abbott, whom you know some little about, has become a member of the City Philosophical Society, which is held at Tatum's house every Wednesday evening. He (Abbott) has sent me a ticket for admission next Wednesday to a lecture; but as you know their rules, I have no need to enter further into them.

'With Abbott I continue a very intimate and pleasing acquaintance. I find him to be a very well-informed young man. His ideas are correct, and his knowledge, general as well as philosophical, is extensive. He acts too with a propriety of behaviour equal to your own, and I congratulate myself much on the acquisition of two such friends as yourself and him.

'How are you situated now? Do you intend to stop in the country, or are you again coming up to London? I was in hopes that I should see you shortly again. Not that I wish to interfere in your arrangements, but for the pleasure it would give me. But I must not be selfish. It is possible that you may be settled where you are at present, or other strong and urgent reasons may exist that will keep you there. If it is so, I wish they may be such as will afford you pleasure, and tend to increase the happiness and comfort of your life.

'I am at present in very low spirits, and scarce know how to continue on in a strain that will be anyway agreeable to you; I will therefore draw to a close this dull epistle, and conclude with wishing you all health and happiness, assuring you that I am sincerely yours,

'M. FARADAY.

'Mr. T. Huxtable, at Mr. Anthony's,
South Moulton, Devonshire.'

Among the scanty notes left by Faraday of his own life, he says: 'Under the encouragement of Mr. Dance,' who had taken him to the lectures at the Royal Institution, 'I wrote to Sir Humphry Davy, sending, as a proof of my earnestness, the notes I had taken of his last four lectures. The reply was immediate, kind, and favourable. After this I continued to work as a book-binder, with the exception of some days during which I was writing as an amanuensis for Sir H. Davy, at the time when the latter was wounded in the eye from an explosion of the chloride of nitrogen.'

Faraday gave to Dr. Paris a fuller account of his first acquaintance with Sir H. Davy. This was published in 'The Life of Davy,' by Dr. Paris, vol. ii. p. 2.

TO J. A. PARIS, M.D.

'Royal Institution, December 23, 1829.

'My dear Sir,—You asked me to give you an account of my first introduction to Sir H. Davy, which I am very happy to do, as I think the circumstances will bear testimony to the goodness of his heart.

'When I was a bookseller's apprentice I was very fond of experiment and very adverse to trade. It happened that a gentleman, a member of the Royal Institution, took me to hear some of Sir H. Davy's last lectures in Albemarle Street. I took notes, and afterwards wrote them out more fairly in a quarto volume.

'My desire to escape from trade, which I thought vicious and selfish, and to enter into the service of Science, which I imagined made its pursuers amiable and liberal, induced me at last to take the bold and simple step of writing to Sir H. Davy, expressing my wishes, and a hope that if an opportunity came in his way he would favour my views; at the same time, I sent the notes I had taken of his lectures.

'The answer, which makes all the point of my communication, I send you in the original, requesting you to take great care of it, and to let me have it back, for you may imagine how much I value it.

1812. ÆT. 21.

'You will observe that this took place at the end of the year 1812; and early in 1813 he requested to see me, and told me of the situation of assistant in the laboratory of the Royal Institution, then just vacant.

'At the same time that he thus gratified my desires as to scientific employment, he still advised me not to give up the prospects I had before me, telling me that Science was a harsh mistress, and in a pecuniary point of view but poorly rewarding those who devoted themselves to her service. He smiled at my notion of the superior moral feelings of philosophic men, and said he would leave me to the experience of a few years to set me right on that matter.

'Finally, through his good efforts, I went to the Royal Institution, early in March of 1813, as assistant in the laboratory; and in October of the same year went with him abroad, as his assistant in experiments and in writing. I returned with him in April 1815, resumed my station in the Royal Institution, and have, as you know, ever since remained there.

'I am, dear Sir, ever truly yours,

'M. FARADAY.'

The following is the note of Sir H. Davy alluded to in Mr. Faraday's letter:—

TO MR. FARADAY.

'December 24, 1812.

'Sir,—I am far from displeased with the proof you have given me of your confidence, and which displays great zeal, power of memory, and attention. I am obliged to go out of town, and shall not be settled in town till the end of January; I will then see you at any time you wish. It would gratify me to be of any service to you; I wish it may be in my power.

'I am, Sir, your obedient humble servant,

'H. DAVY.'

Not only did Sir H. Davy, at his first interview,[1] advise him to keep in business as a bookbinder, but he promised to give him the work of the Institution, as well as his own and that of as many of his friends as he could influence.

One night, when undressing in Weymouth Street, Faraday was startled by a loud knock at the door; and on looking out he saw a carriage from which the footman had alighted and left a note for him. This was a request from Sir H. Davy that he would call on him the next morning. Sir H. Davy then referred to their former interview, and inquired whether he was still in the same mind, telling him that if so he would give him the place of assistant in the laboratory of the Royal Institution, from which he had on the previous day ejected its former occupant. The salary was to be 25*s.* a week, with two rooms at the top of the house.

In the minutes of the meeting of managers on March 1, 1813, is this entry:—'Sir Humphry Davy has the honour to inform the managers that he has found a person who is desirous to occupy the situation in the Institution lately filled by William Payne. His name is Michael Faraday. He is a youth of twenty-two years of age. As far as Sir H. Davy has been able to observe or ascertain, he appears well fitted for the situation. His habits seem good, his disposition active and

[1] This interview took place in the anteroom to the theatre, by the window which is nearest to the corridor.

cheerful, and his manner intelligent. He is willing to engage himself on the same terms as those given to Mr. Payne at the time of quitting the Institution.

'Resolved,—That Michael Faraday be engaged to fill the situation lately occupied by Mr. Payne on the same terms.'

Amongst the few notes he made of his own life there are two or three which relate to 1813. The first has reference to his joining the City Philosophical Society. 'This,' he says, 'was founded in 1808 at Mr. Tatum's house, and I believe by him. He introduced me as a member of the Society in 1813. Magrath was Secretary to the Society. It consisted of thirty or forty individuals, perhaps all in the humble or moderate rank of life. Those persons met every Wednesday evening for mutual instruction. Every other Wednesday the members were alone, and considered and discussed such questions as were brought forward by each in turn. On the intervening Wednesday evenings friends also of the members were admitted, and a lecture was delivered, literary or philosophical, each member taking the duty, if possible, in turn (or in default paying a fine of half a guinea). This society was very moderate in its pretensions, and most valuable to the members in its results.' ('I remember, too,' says one of the members, 'we had a "class-book" in which, in rotation, we wrote essays, and passed it to each other's houses.')

Another note also relates to the self-education which now as ever he was striving to obtain.

'During this spring Magrath and I established the mutual-improvement plan, and met at my rooms up in the attics of the Royal Institution, or at Wood Street at his warehouse. It consisted perhaps of half-a-dozen persons, chiefly from the City Philosophical Society, who met of an evening to read together, and to criticise, correct, and improve each other's pronunciation and construction of language. The discipline was very sturdy, the remarks very plain and open, and the results most valuable. This continued for several years.' Saturday night was the time of meeting at the Royal Institution, in the furthest and uppermost room in the house, then Faraday's place of residence.

The letters which Faraday wrote to Abbott this year give not only an insight into his mind when he first came to the Royal Institution, but show the work on which he immediately entered in the laboratory, and the amount of skill in chemical manipulation which he must have gained by experiments in Blandford Street.

In four of the letters he made remarks on lecture rooms, lectures, apparatus, diagrams, experiments, audiences: these show the keenness of his observation, the abundance of his ideas, and the soundness of his judgment; and it is worthy of notice that he wrote without the slightest forecast of his future career. He says, 'It may perhaps appear singular and improper that one who is entirely unfit for such an office himself, and who does not even pretend to any of the requisites for it, should take upon him to censure and to commend others,' &c.; and yet within two years and a half he began a course of lectures on Chemistry at the City Philosophical Society, and he continued to lecture for thirty-eight years at the Royal Institution. Moreover, the reputation he had with the world in general as a scientific lecturer was certainly not less than that which he gained among scientific men as a philosopher and as an experimentalist. He used every aid to improve his

1813. ÆT. 21.

language and method, and to avoid even the slightest peculiarity; and yet he kept his simplicity and natural manner, as though he had never profited by professional instruction nor owed anything to friendly correction.

As early as March 8, seven days after his appointment, Faraday dates his first letter from the Royal Institution to his friend Abbott.

'Royal Institution, March 8, 1813.

'It is now about nine o'clock, and the thought strikes me that the tongues are going both at Tatum's and at the lecture in Bedford Street; but I fancy myself much better employed than I should have been at the lecture at either of those places. Indeed, I have heard one lecture already to-day, and had a finger in it (I can't say a hand, for I did very little). It was by Mr. Powell, on mechanics, or rather on rotatory motion, and was a pretty good lecture, but not very fully attended.

'As I know you will feel a pleasure in hearing in what I have been or shall be occupied, I will inform you that I have been employed to-day, in part, in extracting the sugar from a portion of beetroot, and also in making a compound of sulphur and carbon—a combination which has lately occupied in a considerable degree the attention of chemists.

'With respect to next Wednesday, I shall be occupied until late in the afternoon by Sir H. Davy, and must therefore decline seeing you at that time; this I am the more ready to do as I shall enjoy your company next Sunday, and hope to possess it often in a short time.

'You must not expect a long letter from me at this time, for I assure you my hand feels somewhat strange in the occupation, and my thoughts come but lazily;

this must plead in excuse for so uninteresting *a compound*, and I entertain but little doubt that it will gain it. 'M. FARADAY.'

His next letter to his friend was written six weeks after he came to the Institution.

'Thursday evening, April 9, 1813.

. . . . . .

'I shall at this time proceed to acquaint you with the results of some more experiments on the detonating compound of chlorine and azote; and I am happy to say I do it at my ease, for I have escaped (not quite unhurt) from four different and strong explosions of the substance. Of these the most terrible was when I was holding between my thumb and finger a small tube containing $7\frac{1}{2}$ grains of it. My face was within twelve inches of the tube; but I fortunately had on a glass mask. It exploded by the slight heat of a small piece of cement that touched the glass above half an inch from the substance, and on the outside. The explosion was so rapid as to blow my hand open, tear off a part of one nail, and has made my fingers so sore that I cannot yet use them easily. The pieces of tube were projected with such force as to cut the glass face of the mask I had on. But to proceed with an account of the experiments:—

'A tube was filled with dry boiled mercury, and inverted in a glass containing also mercury, a portion of the compound was thrown up into it, and it was then left to act all last night. On examining it this morning the compound was gone; a substance was formed in the tube, and a gas obtained: this gas was azote, the substance corrosive mercury, evidently proving it to be a compound of chlorine or oxymuriatic acid gas and azote. On repeating the experiment this morning, as soon as it was thrown up it exploded, and the tube and a receiver were blown to pieces. I got a cut on my eyelid, and Sir H. bruised his hand.

'A portion of it was then introduced into a tube, and a stop-cock connected to it. It was then taken to the air-pump, and exhausted until we supposed the substance to have rose and filled the tube with vapour. It was then heated by a spirit-lamp, and in a few moments an inflammation took place in the tube: but all stood firm. On taking it off from the pump, in order to ascertain the products, it was found that so much common air had passed in from the barrels of the pump as to render the experiment indecisive; and therefore it was repeated this morning with a larger portion of the substance. When put in the pump it was exhausted, and there stood for a moment or two, and then exploded with a fearful noise: both Sir H. and I had masks on, but I escaped this time the best. Sir H. had his face cut in two places about the chin, and a violent blow on the forehead struck through a considerable thickness of silk and leather; and with this experiment he has for the present concluded. The specific gravity of this substance, as ascertained yesterday by comparing its weight with the weight of an equal bulk of water, is 1·95, so that my former estimate is incorrect; but you will excuse it as being the estimate of a tyro in chemical science.

'Such are some few of the properties of this terrible compound, and such are the experiments in which those properties are evinced: from these it appears to be a compound of chlorine and azote, for the presence of any other body has not been satisfactorily proved. It is a body which confers considerable importance on azote, which has till now been celebrated chiefly for negative properties. It shows its energy when united in this combination; and in this compound, too, azote is rendered capable of decomposing the muriatic acid, as shown by the experiment related in my last: it combines with the hydrogen to form ammonia, and the chlorine of the compound and of the acid are liberated.

'This compound is of such explosive power as to render it imprudent to consider it at any time and in any state as secure. Oftentimes it will explode in an experiment that has been before made five or six times with perfect safety, and in which you have been lulled into a dangerous security. I was yesterday putting some into a clean dry tube, when it exploded on touching the glass, and rushed in my face; so that it is, as I before said, improper to consider it at any time as secure.

. . . . . .

'But away with philosophy at present. Remember me to all friends within the ethereal atmosphere of Bermondsey; and believe me to be, what I hope shortly to assure you personally I am, yours truly,

'M. FARADAY.'

The next letter to Abbott is written on May 12.

'The monk, for the chastisement of his body and mortification of his sensual lusts and worldly appetites, abstains from pleasures and even the simple supplies that nature calls for; the miser, for reasons as strong though diametrically opposite—the gratification of a darling passion—does exactly the same, and leaves unenjoyed every comfort of life; but I, for no reason at all, have neglected that which constitutes one of my greatest pleasures, and one that may be enjoyed with the greatest propriety, till on a sudden, as the dense light of the electric flash pervades the horizon, so struck the thought of A. through my soul.

'And yet, B., though I mean to write to you at this time, I have no subject in particular out of which I can cut a letter. I shall, therefore (if you will allow me a second simile), follow the pattern of the expert sempstress who, when she has cut out all her large and important works, collects and combines, as fancy may direct, pieces of all sorts and sizes, shapes and colours, and calls it patch-work—such a thing will this epistle most probably turn out; begun one day, yet most likely finished on another; formed of things no otherwise connected than as they stand upon the paper—things, too, of different kinds. It may well be called patch-work, or work which pleases none more than the maker. What is the matter with the thumb and forefinger of your right hand? and yet, though they be ever so much out of order, it can scarcely excuse your long silence. I have expected something from you before now, even though it might be written with the left hand.

'"He that hath not music in his heart," &c. Confound the music, say I; it turns my thoughts quite round, or rather half way round, from the letter. You must know, Sir, that there is a grand party at dinner at Jacques' Hotel, which immediately faces the back of the Institution; and the music is so excellent, that I cannot for the life of me help running at every new piece they play to the window to hear them. I shall do no good at this letter to-night, and so will get to bed, and "listen, listen to the voice of" bassoons, violins, clarionettes, trumpets, serpents, and all the accessories to good music. I can't stop. Good-night.

'May 14th.—What a singular compound is man! what strange contradictory ingredients enter into his composition, and how completely each one predominates for a time, according as it is favoured by the tone of the mind and senses, and other exciting circumstances!—at one time grave, circumspect, and cautious; at another, silly, headstrong, and careless;—now conscious of his dignity, he considers himself a lord of the creation, yet in a few hours will conduct himself in a way that places him beneath the level of beasts; at times free, frivolous, and open, his tongue is an unobstructed conveyer of his thoughts—thoughts which, on after-consideration, make him ashamed of his former behaviour; indeed, the numerous paradoxes, anomalies, and contradictions in man exceed in number all that can be found in nature elsewhere, and separate and distinguish him, if nothing else did, from every other created object, organised or not. The study of these circumstances is not uninteresting, inasmuch as knowledge of them enables us to conduct ourselves with much more propriety in every situation in life. Without knowing how far we ourselves are affected by them, we should be unable to trust to our discretion amongst other persons; and without some knowledge of the part they bear or make in their own position, we should be unable to behave to them unreserved and with freedom.

'It was my intention, when I again sat down to this letter, to obliterate all the former part of it; but the thoughts I have just set down were sufficient to alter my determination. I have left them as being the free utterance of an unemployed mind, and delineating a true part of my constitution; I believe, too, that I know sufficiently of the component parts of my friend to justify my confidence in letting them remain unaltered.

'For much more I have neither room nor time to spare; nor, had I, would I lengthen what is already too long; yet, as a clock, after giving warning, passes on for a few moments before it strikes, so do I linger on the paper. It is my intention to accept of your kind invitation for Sunday *morning* (further your deponent knoweth not); and I shall, therefore, take the liberty of seeing you after breakfast, at about 9.45; till when I remain, with respects to all friends, yours truly,

'M. FARADAY.'

The next letter to Abbott is dated June 1, 1813.

'Dear A——, Again I resort, for pleasure and to dispel the dulness of a violent headache, to my correspondence with you, though perfectly unfit for it except as it may answer the purpose of amusing myself. The subject upon which I shall dwell more particularly at present has been in my head for some considerable time, and it now bursts forth in all its confusion. The opportunities that I have latterly had of attending and obtaining instruction from various lecturers in their performance of the duty attached to that office, has enabled me to observe the various habits, peculiarities, excellences, and defects of each of them as they were evident to me during the delivery. I did not wholly let this part of the things occurrent escape my notice, but when I found myself pleased, endeavoured to

1813.
ÆT. 21.

ascertain the particular circumstance that had affected me; also, whilst attending Mr. Brande and Mr. Powell in their lectures, I observed how the audience were affected, and by what their pleasure and their censure were drawn forth.

'It may, perhaps, appear singular and improper that one who is entirely unfit for such an office himself, and who does not even pretend to any of the requisites for it, should take upon him to censure and to commend others, to express satisfaction at this, to be displeased with that, according as he is led by his judgment, when he allows that his judgment is unfit for it; but I do not see, on consideration, that the impropriety is so great. If I am unfit for it, 'tis evident that I have yet to learn, and how learn better than by the observation of others? If we never judge at all, we shall never judge right; and it is far better to learn to use our mental powers (though it may take a whole life for the purpose) than to leave them buried in idleness a mere void.

'I too have inducements in the C.P.S. (City Philosophical Society) to draw me forward in the acquisition of a small portion of knowledge on this point, and these alone would be sufficient to urge me forward in my judgment of $\left\{\begin{array}{l}\text{men}\\ \text{lectures}\end{array}\right\}$ and $\left\{\begin{array}{l}\text{things}\\ \text{apparatus.}\end{array}\right.$

'In a word, B., I intend to give you my ideas on the subject of lectures and lecturers in general. The observations and ideas I shall set down are such as entered my mind at the moment the circumstances that gave rise to them took place. I shall point out but few beauties or few faults that I have not witnessed in the presence of a numerous assembly; and it is exceedingly probable, or rather certain, that I should have noticed more of these particulars if I had seen more lecturers; or, in other words, I do not pretend to give you an account of all the faults possible in a lecture, or directions for the composing and delivering of a perfect one.

'On going to a lecture I generally get there before it begins; indeed, I consider it as an impropriety of no small magnitude to disturb the attention of an audience by entering amongst them in the midst of a lecture, and, indeed, bordering on an insult to the lecturer. By arriving there before the commencement, I have avoided this error, and have had time to observe the lecture room.'

(He then dwells on the form of the lecture room.)

'There is another circumstance to be considered with respect to a lecture room of as much importance almost as light itself, and that is ventilation. How often have I felt oppression in the highest degree when surrounded by a number of other persons, and confined in one portion of air! How have I wished the lecture finished, the lights extinguished, and myself away merely to obtain a fresh supply of that element! The want of it caused the want of attention, of pleasure, and even of comfort, and not to be regained without its previous admission. Attention to this is more particularly necessary in a lecture room intended for night delivery, as the lights burning add considerably to the oppression produced on the body.

'Entrance and exit are things, too, worthy of consideration amongst the particulars of a lecture room; but I shall say no more on them than to refer you to the mode in which this is arranged here—a mode excellently well adapted for the convenience of a great number of persons.

'Having thus thrown off, in a cursory manner, such thoughts as spontaneously entered my mind on this part of the subject, it appears proper next to consider the subjects fit for the purposes of a lecture. Science is undeniably the most eminent in its fitness for this purpose. There is no part of it that may not be treated of, illustrated, and explained with profit and pleasure to the hearers in this manner. The facility, too, with which it allows of manual and experimental illustration places it foremost in this class of subjects. After it come (as I conceive) arts and manufactures, the polite arts, belles lettres, and a list which may be extended until it includes almost every thought and idea in the mind of man, politics excepted. I was going to add religion to the exception, but remembered that it is explained and laid forth in the most popular and eminent manner in this way. The fitness of subjects, however, is connected in an inseparable manner with the kind of audience that is to be present, since excellent lectures in themselves would appear absurd if delivered before an audience that did not understand them. Anatomy would not do for the generality of audiences at the Royal Institution, neither would metaphysics engage the attention of a company of schoolboys. Let the subject fit the audience, or otherwise success may be despaired of.

'A lecturer may consider his audience as being polite or vulgar (terms I wish you to understand according to Shuffleton's new dictionary), learned or unlearned (with respect to the subject), listeners or gazers. Polite company expect to be entertained not only by the subject of the lecture, but by the manner of the lecturer; they look for respect, for language consonant to their dignity, and ideas on a level with their own. The vulgar—that is to say in general, those who will take the trouble of thinking, and the bees of business—wish for something that they can comprehend. This may be deep and elaborate for the learned, but for those who are as yet tyros and unacquainted with the subject must be simple and plain. Lastly, listeners expect reason and sense, whilst gazers only require a succession of words.

'These considerations should all of them engage the attention of the lecturer whilst preparing for his occupation, each particular having an influence on his arrangements proportionate to the nature of the company he expects. He should consider them connectedly, so as to keep engaged completely during the whole of the lecture the attention of his audience.

'If agreeable, this subject shall be resumed at a future time; till when I am, as always, yours sincerely,

'M. FARADAY.'

The next letter to Abbott is dated June 4th.

. . . . . .

'I need not point out to the active mind of my friend the astonishing disproportion, or rather difference, in the perceptive powers of the eye and the ear, and the facility and clearness with which the first of these organs conveys ideas to the mind—ideas which, being thus gained, are held far more retentively and firmly in the memory than when introduced by the ear. 'Tis true the ear here labours under a disadvantage, which is that the lecturer may not always be qualified to state a fact with the utmost precision and clearness that language allows him and that the ear can understand, and thus the complete action of the organ, or rather of its assigned portion of the sensorium, is not called forth; but this evidently points out to us the necessity of aiding it by using the eye also as a medium for the attainment of knowledge, and strikingly shows the necessity of apparatus.

'Apparatus therefore is an essential part of every lecture in which it can be introduced; but to apparatus should be added, at every convenient opportunity, illustrations that may not perhaps deserve the name of apparatus and of experiments, and yet may be introduced with considerable force and effect in proper places. Diagrams, and tables too, are necessary, or at least add in an eminent degree to the illustration and perfection of a lecture. When an experimental lecture is to be delivered, and apparatus is to be exhibited, some kind of order should be observed in the arrangement of them on the lecture table. Every particular part illustrative of the lecture should be in view, no one thing should hide another from the audience, nor should anything stand in the way of or obstruct the lecturer. They should be so placed, too, as to produce a kind of uniformity in appearance. No one part should appear naked and another crowded, unless some particular reason exists and makes it necessary to be so. At the same time, the whole should be so arranged as to keep one operation from interfering with another. If the lecture table appears crowded, if the lecturer (hid by his apparatus) is invisible, if things appear crooked, or aside, or unequal, or if some are out of sight, and this without any particular reason, the lecturer is considered (and with reason too) as an awkward contriver and a bungler.'

(He then dwells on diagrams and illustrations.)

'June 5, six o'clock P.M.

'I have but just got your letter, or should have answered it before. For your request—it is fulfilled; for your invitation—I thank you, but cannot accept it; for your orders—they shall be attended to; for to see you—I will come on Tuesday evening; and for want of time—I must conclude, with respects to all friends, yours sincerely,

'M. FARADAY.'

Again, June 11, he writes to Abbott.

'The most prominent requisite to a lecturer, though perhaps not really the most important, is a good delivery; for though to all true philosophers science and nature will have charms innumerable in every dress, yet I am sorry to say that the generality of mankind cannot accompany us one short hour unless the path is strewed with flowers. In order, therefore, to gain the attention of an audience (and what can be more disagreeable to a lecturer than the want of it?), it is necessary to pay some attention to the manner of expression. The utterance should not be rapid and hurried, and consequently unintelligible, but slow and deliberate, conveying ideas with ease from the lecturer, and infusing them with clearness and readiness into the minds of the audience. A lecturer should endeavour by all means to obtain a facility of utterance, and the power of clothing his thoughts and ideas in language smooth and harmonious and at the same time simple and easy. His periods should be round, not too long or unequal; they should be complete and expressive, conveying clearly the whole of the ideas intended to be conveyed. If they are long, or obscure, or incomplete, they give rise to a degree of labour in the minds of the hearers which quickly causes lassitude, indifference, and even disgust.

'With respect to the action of the lecturer, it is requisite that he should have some, though it does not here bear the importance that it does in other branches of oratory; for though I know of no species of delivery (divinity excepted) that requires less motion, yet I would by no means have a lecturer glued to the table or screwed on the floor. He must by all means appear as a body distinct and separate from the things around him, and must have some motion apart from that which they possess.

'A lecturer should appear easy and collected, undaunted and unconcerned, his thoughts about him, and his mind clear and free for the contemplation and description of his subject. His action should not be hasty and violent, but slow, easy, and natural, consisting principally in changes of the posture of the body, in order to avoid the air of stiffness or sameness that would otherwise be unavoidable. His whole behaviour should evince respect for his audience, and he should in no case forget that he is in their presence. No accident that does not interfere with their convenience should disturb his serenity, or cause variation in his behaviour; he should never, if possible, turn his back on them, but should give them full reason to believe that all his powers have been exerted for their pleasure and instruction.

'Some lecturers choose to express their thoughts

extemporaneously immediately as they occur to the mind, whilst others previously arrange them, and draw them forth on paper. Those who are of the first description are certainly more unengaged, and more at liberty to attend to other points of delivery than their pages; but as every person on whom the duty falls is not equally competent for the prompt clothing and utterance of his matter, it becomes necessary that the second method should be resorted to. This mode, too, has its advantages, inasmuch as more time is allowed for the arrangement of the subject, and more attention can be paid to the neatness of expression.

'But although I allow a lecturer to write out his matter, I do not approve of his reading it; at least, not as he would a quotation or extract. He should deliver it in a ready and free manner, referring to his book merely as he would to copious notes, and not confining his tongue to the exact path there delineated, but digress as circumstances may demand or localities allow.

'A lecturer should exert his utmost effort to gain completely the mind and attention of his audience, and irresistibly to make them join in his ideas to the end of the subject. He should endeavour to raise their interest at the commencement of the lecture, and by a series of imperceptible gradations, unnoticed by the company, keep it alive as long as the subject demands it. No breaks or digressions foreign to the purpose should have a place in the circumstances of the evening; no opportunity should be allowed to the audience in which their minds could wander from the subject, or return to inattention and carelessness.

A flame should be lighted at the commencement, and kept alive with unremitting splendour to the end. For this reason I very much disapprove of breaks in a lecture, and where they can by any means be avoided, they should on no account find place. If it is unavoidably necessary, to complete the arrangement of some experiment, or for other reasons, leave some experiments in a state of progression, or state some peculiar circumstance, to employ as much as possible the minds of the audience during the unoccupied space—but, if possible, avoid it.

'Digressions and wanderings produce more or less the bad effects of a complete break or delay in a lecture, and should therefore never be allowed except in very peculiar circumstances; they take the audience from the main subject, and you then have the labour of bringing them back again (if possible).

'For the same reason (namely, that the audience should not grow tired), I disapprove of long lectures; one hour is long enough for anyone, nor should they be allowed to exceed that time.

'But I have said enough for once on this subject, and must leave it in order to have room for other things. I had arranged matters so as to accept your kind invitation for Sunday, and anticipated much pleasure from the meeting, but am disagreeably disappointed, circumstances being such as to hinder my seeing you at that time. This I much regret, but hope, however, to enjoy the full measure of pleasures expected at some not far distant time.

'But farewell, dear A., for a few days, when you shall again hear from yours most sincerely,

'M. FARADAY.'

1813.
ÆT. 21.

The last letter to Abbott before he went abroad is dated June 18, 1813.

'Dear A——, As when on some secluded branch in forest far and wide sits perched an owl, who, full of self-conceit and self-created wisdom, explains, comments, condemns, ordains, and orders things not understood, yet full of his importance still holds forth to stocks and stones around—so sits and scribbles Mike; so he declaims to walls, stones, tables, chairs, hats, books, pens, shoes, and all the things inert that be around him, and so he will to the end of the chapter.

'In compliance with that precept which desires us to finish one thing before we begin another, I shall at once fall to work on the lecturer, and continue those observations which I have from time to time both made and gained about them. Happy am I to say that the fault I shall now notice has seldom met my observation, yet, as I have witnessed it, and as it does exist, it is necessary to notice it.

'A lecturer falls deeply beneath the dignity of his character when he descends so low as to angle for claps, and asks for commendation. Yet have I seen a lecturer even at this point. I have heard him causelessly condemn his own powers. I have heard him dwell for a length of time on the extreme care and niceness that the experiment he will make requires. I have heard him hope for indulgence when no indulgence was wanted, and I have even heard him declare that the experiment now made cannot fail from its beauty, its correctness, and its application, to gain the approbation of all. Yet surely such an error in the character of a lecturer cannot require pointing out, even to those who resort to it; its impropriety must be evident, and I should perhaps have done well to pass it.

'Before, however, I quite leave this part of my subject, I would wish to notice a point in some manner connected with it. In lectures, and more particularly experimental ones, it will at times happen that accidents or other incommoding circumstances take place. On these occasions an apology is sometimes necessary, but not always. I would wish apologies to be made as seldom as possible, and generally, only when the inconvenience extends to the company. I have several times seen the attention of by far the greater part of the audience called to an error by the apology that followed it.

'An experimental lecturer should attend very carefully to the choice he may make of experiments for the illustration of his subject. They should be important, as they respect the science they are applied to, yet clear, and such as may easily and generally be understood. They should rather approach to simplicity, and explain the established principles of the subject, than be elaborate, and apply to minute phenomena only. I speak here (be it understood) of those lectures which are delivered before a mixed audience, and the nature of which will not admit of their being applied to the explanation of any but the principal parts of a science. If to a particular audience you dwell on a particular subject, still adhere to the same principle, though perhaps not exactly to the same rule. Let your experiments apply to the subject you elucidate, do not introduce those which are not to the point.

'Though this last part of my letter may appear

superfluous, seeing that the principle is so evident to every capacity, yet I assure you, dear A., I have seen it broken through in the most violent manner—a mere alehouse trick has more than once been introduced in a lecture, delivered not far from Pall Mall, as an elucidation of the laws of motion.

'Neither should too much stress be laid upon what I would call small experiments, or rather illustrations. It pleases me well to observe a neat idea enter the head of a lecturer, the which he will immediately and aptly illustrate or explain by a few motions of his hand—a card, a lamp, a glass of water, or any other thing that may be by him; but when he calls your attention in a particular way to a decisive experiment that has entered his mind, clear and important in its application to the subject, and then lets fall a card, I turn with disgust from the lecturer and his experiments. 'Tis well, too, when the lecturer has the ready wit and the presence of mind to turn any casual circumstance to an illustration of his subject. Any particular circumstance that has become table-talk for the town, any local advantages or disadvantages, any trivial circumstance that may arise in company, give great force to illustrations aptly drawn from them, and please the audience highly, as they conceive they perfectly understand them.

'Apt experiments (to which I have before referred) ought to be explained by satisfactory theory, or otherwise we merely patch an old coat with new cloth, and the whole (hole) becomes worse. If a satisfactory theory can be given, it ought to be given. If we doubt a received opinion, let us not leave the doubt unnoticed, and affirm our own ideas, but state it clearly, and lay down also our objections. If the scientific world is divided in opinion, state both sides of the question, and let each one judge for himself, by noticing the most striking and forcible circumstances on each side. Then, and then only, shall we do justice to the subject, please the audience, and satisfy our honour, the honour of a philosopher. I shall here cause a slight separation in the subject by closing this epistle, as it is now getting late; so I shake hands until to-morrow, at which time I hope to find all well, as is at present

'Yours sincerely,

'M. FARADAY.'

When urged by his friend, two years after this, to complete his remarks, he said, December 31, 1816, 'With respect to my remarks on lectures, I perceive I am but a mere tyro in the art, and therefore you must be satisfied with what you have, or expect at some future time a recapitulation, or rather revision, of them.'

The short history of himself which he gave in a letter written on September 13, 1813, to his aunt and uncle, because 'he has nothing more to say, and is requested by his mother to write the account,' is highly characteristic of the man.

'I was formerly a bookseller and binder, but am now turned philosopher, which happened thus:—Whilst an apprentice, I, for amusement, learnt a little of chemistry and other parts of philosophy, and felt an eager desire to proceed in that way further. After being a journeyman for six months, under a disagreeable master, I gave up my business, and, by the interest of Sir H. Davy, filled the situation of chemical assistant to the Royal Institution of Great Britain, in which office I now remain, and where I am constantly engaged in observing the works of Nature and tracing the manner in which she directs the arrangement and order of the world. I have lately had proposals made to me by Sir Humphry Davy to accompany him, in his travels through Europe and into Asia, as philosophical assistant. If I go at all I expect it will be in October next, about the end, and my absence from home will perhaps be as long as three years. But as yet all is uncertain. I have to repeat that, even though I may go, my path will not pass near any of my relations, or permit me to see those whom I so much long to see.'

In his notes he says:—'In the autumn Sir H. Davy proposed going abroad, and offered me the opportunity of going with him as his amanuensis, and the promise of resuming my situation in the Institution upon my return to England. Whereupon I accepted the offer, left the Institution on October 13, and, after being with Sir H. Davy in France, Italy, Switzerland, the Tyrol, Geneva, &c., in that and the following year, returned to England and London April 23, 1815.'

CHAPTER III.

EXTRACTS FROM HIS JOURNAL AND LETTERS WHILST ABROAD WITH SIR HUMPHRY DAVY.

1813. ÆT. 22.

THE journey of Faraday abroad with Sir H. Davy was one of the few episodes that occurred in his life. It lasted only one year and a half.

During this time he kept a journal, and wrote letters to his mother, sisters, and friends; chiefly, however, to Abbott.

The journal, of which only some extracts are here given, is remarkable for the minuteness of the description of all he saw, and for its cautious silence regarding those he was with. It gives, however, full details regarding Sir H. Davy's scientific work. He says he wrote it, 'not to instruct or to inform, or to convey even an imperfect idea of what it speaks; its sole use is to recall to my mind at some future time the things I see now, and the most effectual way to do that will be, I conceive, to write down, be they good or bad, my present impressions.'

The letters are full of the warmth of his affection, the sensitiveness of his feeling, and the earnestness of his desire for self-improvement.

In one of his first letters to his mother he says, 'The first and last thing in my mind is England, home, and

1813. ÆT. 22.

friends. It is the point to which my thoughts still ultimately tend, the goal to which, looking over intermediate things, my eyes are still directed. . . Whenever a vacant hour occurs, I employ it by thinking of those at home. Whenever present circumstances are disagreeable, I amuse myself by thinking of those at home. In short, when sick, when cold, when tired, the thoughts of those at home are a warm and refreshing balm to my heart . . . these are the first and greatest sweetness in the life of man.'

His desire for improvement is seen in another letter which he writes later to his mother: 'I am almost contented except with my ignorance, which becomes more visible to me every day, though I endeavour as much as possible to avoid it.'

On this subject also he writes to his friend, 'I have several times been more than half decided to return hastily home, but second thoughts have still induced me to try what the future may produce, and now I am only retained by the wish of improvement. I have learned just enough to perceive my ignorance, and, ashamed of my defects in everything, I wish to seize the opportunity of remedying them. The little knowledge I have gained in languages makes me wish to know more of them, and the little I have seen of men and manners is just enough to make me desirous of seeing more. Added to which, the glorious opportunity I enjoy of improving in the knowledge of chemistry and the sciences continually determines me to finish this voyage with Sir Humphry Davy.'

To his married sister he thus shows his affection: 'I shall never feel quite happy until I get amongst you again. I have a thousand things to say, but I do not know which to say first; and if I followed my mind, I should never get to an end.'

And in his last letter to his mother from Brussels, he says, 'I have a thousand times endeavoured to fancy a meeting with you and my relations and friends, and I am sure I have as often failed—the reality must be a pleasure not to be imagined nor to be described . . . My thoughts wander from one to another, my pen runs on by fits and starts . . . I do not know what to say, and yet I cannot put an end to my letter. I would fain be talking to you, but I must cease . . . It is the shortest, and to me the sweetest, letter I ever wrote you.'

Faraday began his foreign journal thus:—

Wednesday, October 13*th.*—This morning formed a new epoch in my life. I have never before, within my recollection, left London at a greater distance than twelve miles; and now I leave it perhaps for many years, and to visit spots between which and home whole realms will intervene. 'Tis indeed a strange venture at this time, to trust ourselves in a foreign and hostile country, where also so little regard is had to protestations and honour, that the slightest suspicion would be sufficient to separate us for ever from England, and perhaps from life. But curiosity has frequently incurred dangers as great as these, and therefore why should I wonder at it in the present instance? If we return safe, the pleasures of recollection will be highly enhanced by the dangers encountered; and a never-failing consolation is, that whatever be the fate of our party, variety, a great source of amusement and pleasure, must occur.

1813. ÆT. 22.

Friday, 15*th*.—Reached Plymouth this afternoon. I was more taken by the scenery to-day than by anything else I have ever seen. It came upon me unexpectedly, and caused a kind of revolution in my ideas respecting the nature of the earth's surface. That such a revolution was necessary is, I confess, not much to my credit; and yet I can assign to myself a very satisfactory reason, in the habit of ideas induced by an acquaintance with no other green surface than that within three miles of London. Devonshire, however, presented scenery very different to this; the mountainous nature of the country continually put forward new forms and objects, and the landscape changed before the eye more rapidly than the organ could observe it. This day gave me some ideas of the pleasures of travelling, and has raised my expectations of future enjoyment to a very high point.

Monday, 18*th*.—I last night had a fine opportunity of observing the luminous appearance of the sea, and was amused by it for a long time. As the prow of the vessel met the waters, it seemed to turn up a vast number of luminous bodies about the size of peas, some, however, being larger than others. These appeared to roll onwards by the side of the vessel with the waters, and sometimes traversed a distance of many yards before they disappeared. They were luminous at or beneath the surface of the water indifferently, and the only effect produced by different depths was a diminution of the light by the quantity of intervening medium. These luminous spots were very numerous —the most so, I think, about half an hour after midnight: their light was very bright and clear.

The swell of the sea was very considerable all night,

though gradually decreasing. I remained on deck and escaped all sea-sickness. As day came on and the light increased, we looked about us, but saw nothing in the scene except sky and immense waves striding one after the other at a considerable distance. These as they came to us lifted up our small vessel, and gave us, when on their summits, a very extended horizon: but we soon sank down into the valleys between them, and had nothing in view but the wall of waters around us.

Tuesday 19*th*.—As soon as day was well introduced our vessel moved, and, passing the cartel which stands at the mouth of the harbour (of Morlaix) to defend it, moved up the long and perplexed passage. We here had our first view of France, and it was not at all calculated to impress a stranger with a high opinion of the country, though perhaps regret for home may influence first feelings. I was in hopes of going on shore, but understood that no one could leave the ship until the arrival of an officer to examine us. Late in the afternoon the mighty man of office came, attended by several understrappers and a barge full of Frenchmen, apparently beggars and porters. A formal examination then ensued. One of the officers came to me, and, taking my hat off, he first searched it, and then laid it on the deck; he then felt my pockets, my breast, my sides, my clothes, and, lastly, desired to look into my shoes; after which I was permitted to pass. A similar ceremony was performed on all the strangers; and though I felt surprised at such a singular reception, I could hardly help laughing at the ridiculous nature of their precautions. Our English sailors looked on with pity and indignation, which was not diminished by the seizure of some letters written in the harbour, and given to the captain of the cartel to be conveyed back to Plymouth, and by the post to our friends in London. These letters, however, were all seized and conveyed to Morlaix, and we were not allowed to write home of our arrival in France.

The various parts of the carriage, the boxes, packages, &c., being placed on deck, word was given, and immediately the crew of Frenchmen poured on them, and conveyed them in every direction, and by the most awkward and irregular means, into the barge alongside, and this with such an appearance of hurry and bustle, such an air of business and importance, and yet so ineffectually, that sometimes nine or ten men would be round a thing of a hundred pounds' weight, each most importantly employed; and yet the thing would remain immoveable until the crew were urged by their officer or pushed by the cabin boy. At last all was placed in the barge, and then leave was given to the cartel to return. And certainly it was with no pleasurable feelings I beheld myself separated from my countrymen, that I saw them returning, and felt conscious of the tyrannical and oppressive laws and manners of the people in whose hands we remained. But things being as they were, I endeavoured to content and amuse myself by looking out for variety in the manners of the people round me.

Wednesday 20th.—The officers had permitted us to take out of the seat-boxes, &c., in their presence, what was absolutely necessary for the night, and in the morning we went to claim the rest. I found the carriage, &c., in the barge just as they had been left, and an officer still there. The douane was not yet open, and we had to wait patiently, or otherwise, for some time, looking on our things, but not daring to touch them. At last business commenced. The officers having arranged themselves on the edge of the quay, some thirty or forty inhabitants of the town ran and tumbled down the steps, and leaping into the barge, seized, some one thing, some another, and conveyed them to the landing-place above. This sight alone was a curious one, for they being totally destitute of all method and regularity, it seemed as if a parcel of thieves were scampering away with what was not their own. The body of the carriage was the part which most embarrassed them, for as there were no cranes or any substitute for them on the quay, it was necessary that mere hand labour should perform the important task of raising it to the place above. This was an effort of great magnitude, but they *manfully* surmounted it, and our fears of seeing the carriage resigned to its fate at the bottom of the stairs were fortunately unfounded.

All this being done, these gentry formed a ring, and the officers began their work. All the boxes and packages, even to the tool-chest, were taken out and conveyed into the house; and then, some getting inside and some mounting outside, they searched all the corners and crannies for what they could find, and thumped over every part of the carriage to discover hollow and secret places. Finding nothing like concealment, they entered the house, and began to operate on the trunks; and as they were disappointed in their hopes of booty from the carriage they seemed determined to make up for their loss here. Package after package was opened, roll after roll unfolded, each pair of stockings unwrapped, and each article of apparel shaken; but still being disappointed in their hopes of a pretext for seizure, they at last laid claim to two or three dozen of cotton stockings because they were new, and it was long before the arguments of their being necessary for a long journey, and of their being marked, were sufficient to induce them to render them up again. At last the business ended with everything in the possession of the rightful owners, and a gift to the officers for their *polite* attentions.

As soon as the examination was concluded, leave was given for the carriage to be put together and the goods replaced in it. The first set of men now found work again; and I was astonished how, with their poor means and their want of acquaintance with such affairs, they were still able to get it in order. 'Tis true they made the job appear a mighty one, but they got through it; and after having exclaimed *levez! levez!* for an hour or two everything was in a movable state; and horses being tied to, we proceeded in order to the hotel.

I shall refrain from making comments upon this peculiar examination, except to remark, that if variety be one of the traveller's pleasures, we have certainly enjoyed a very high one this morning, for the whole affair was so different to anything that I had before witnessed that I cannot possibly charge it with a monotonous effect: the occurrence was one which will ever make this day signal in my remembrance.

Thursday, 21*st*.—I will endeavour to describe our hotel. This, the best in the place, has but one entrance, and it is paved in a manner similar to the street: through it pass, indiscriminately, horses, pigs, poultry, human beings, or whatever else has a connection with the house or the stables and pigsties behind it. On the right hand of the passage, and equally public as a thoroughfare with it, is the kitchen: here a fire of wood is generally surrounded by idlers, beggars or nondescripts of the town, who meet to warm themselves and chatter to the mistress; and they hold their stations most tenaciously, though the processes of cooking are in progress. I think it is impossible for an English person to eat the things that come out of this place except through ignorance or actual and oppressive hunger; and yet perhaps appearances may be worse than the reality, for in some cases their dishes are to the taste excellent and inviting, but then they require, whilst on the table, a dismissal of all thoughts respecting the cookery or kitchen.

Friday, 22nd.—The postilion deserves a paragraph to himself. He is mostly a young, always a lively, man. His dress, with the exception of his boots and that part which covers his head, varies infinitely, but hairy jackets appear to be frequent as outer garments, and they are often finely ornamented; at other times the dress seems to be a kind of uniform, being at many posthouses together of one colour, and turned up at the edge with another. The first pair of jack-boots that I saw came out of the kitchen at the hotel at Morlaix; for as it is almost impossible for a man when in them to move about by his own exertions, the postilion had left them in the above-named place until all was arranged at the carriage; but then he used his reserved strength, and showed them off in a walk from the fireside to the horses. They appeared like two very large cylinders of leather terminated at the end by purses for the feet; they rose about six inches above the knee, and were cut away at the back part to admit the use of that joint. Their external diameter was about seven inches, but the cavities within were not much too large for the legs. The sides of the boots consisted of two or three folds of strong leather sewed together, and stuffed on the inside with wool to the thickness of three-quarters of an inch and sometimes more, and the lower part, or foot, not being stuffed in the same way, was much smaller in proportion, though, being still too large, it was made perfect by a wisp of straw. The weight of a pair of jack-boots varies between fourteen and twenty pounds generally. These boots are sometimes moved about by the postilions independent of the exertions of the horses, and then an enormous pair of stirrups are hung to the saddle to sustain them in riding. At other times they are attached to the saddle by straps, and the postilion jumps on to his horse and into them at the same time. The use of them, according to the wearers, is to save their legs from being broken should the horses stumble or the carriage be overturned; and though a traveller must laugh at the sight of such clumsy things, there is not much amusement in the idea that the people who best know their horses and drivers consider such a precaution constantly necessary.

Other appendages to the postilion are the whip and the tobacco pouch. The first is a most tremendous weapon to dogs, pigs, and little children. With a handle of about thirty inches, it has a thong of six to eight feet in length, and it is constantly in a state of violent vibratory motion over the heads of the horses, giving rise to a rapid succession of stunning sounds. The second is generally a bag, though sometimes a pocket, exclusively appropriated, answers the purpose. It contains tobacco, a short pipe, a flint, a steel, German tinder, and sometimes a few varieties. To this the postilion has constant recurrence, and whilst jogging on will light his pipe and smoke it out successively for several hours.

Sunday, 24th.—This evening I for the first time saw a glowworm. The night was very dark (about seven o'clock), and one of our horses had tumbled over. This accident destroyed the traces; and whilst the postilion was renovating them, I saw the little insect by its light among the horses' feet in the middle of the road. Two small luminous spots were visible upon it, but the light was very weak. I picked the worm up, and secured it until we were again in a moving state, and then amused half an hour by observing its appearance. The lights had disappeared, but soon became visible, and then showed a varying intensity for some minutes, but soon entirely disappeared. On examining it afterwards at Rennes, I found it to be a small black worm not three-quarters of an inch in length, and having no part particularly distinguished as that which had been luminous. It was dead, and must have been in a very weak state when I found it.

Thursday, 28th.—Drieux. I cannot help dashing a note of admiration to one thing found in this part of the country—the pigs! At first I was positively doubtful of their nature, for though they have pointed noses, long ears, rope-like tails, and cloven feet, yet who would have imagined that an animal with a long thin body, back and belly arched upwards, lank sides, long slender feet, and capable of outrunning our horses for a mile or two together, could be at all allied to the fat sow of England? When I first saw one, which was at Morlaix, it started so suddenly, and became so active in its motions on being disturbed, and so dissimilar in its actions to our swine, that I looked out for a second creature of the same kind before I ventured to decide on its being a regular animal or an extraordinary production of nature; but I find that they are all alike, and that what at a distance I should judge to be a greyhound I am obliged, on a near approach, to acknowledge a pig.

Friday, 29*th*.—Paris. I am here in the most unlucky and irritating circumstances possible. Set down in the heart of Paris—that spot so desiringly looked after, so vainly too, from a distance by numbers of my countrymen. I know nothing of the language or of a single being here; added to which the people are enemies, and they are vain. My only mode will be to stalk about the town, looking and looked at like a man in the monkish catacombs. My mummies move, however, and they see with their eyes. I must exert myself to attain their language so as to join in their world.

Saturday, 30*th*.—I saw the Galerie Napoléon to-day, but I scarcely know what to say of it. It is both the glory and the disgrace of France. As being itself, and as containing specimens of those things which proclaim the power of man, and which point out the high degree of refinement to which he has risen, it is unsurpassed, unequalled, and must call forth the highest and most unqualified admiration; but when memory brings to mind the manner in which the works came here, and views them only as the gains of violence and rapine, she blushes for the people that even now glory in an act that made them a nation of thieves.

The museum contains paintings, statues, pieces of sculpture and casting, of which by far the greater number have been brought from Italy—they are the works of the old and most eminent masters, and it is a collection of *chefs d'œuvre.* The statues are arranged in the lower part of the Louvre, in many *salons* of great magnificence. There are amongst them the Apollo, the Laocoon, the Venus de' Medici, the Heracles, the Gladiator dying, and many more of the finest pieces of the ancient Greek masters.

Tuesday, Nov. 2nd.—The streets of Paris are paved with equality—that is to say, no difference is made in them between men and beasts, and no part of the street is appropriated to either; add to this that the stones of which the pavement consists are very small and sharp to the foot, and I think much more need not be said in praise of it. At this season, also, besides the pain caused by this sort of pavement, an additional inconvenience arises from it; for though in fine weather a walker may make up his mind to skip across a street half a dozen times in the length of it, to avoid the carriages that drive down upon him, and from which he has no other means of saving himself, yet when, in frosty weather, the sink has become choked up, and the street is overflowed by the never-ceasing fountain, he feels averse to plunge himself into a pond though to save himself from a carriage; and when he does do so, he generally adds energy to the desperation required by an exclamation.

Tuesday, 9th.—I went to-day to La Prefecture de Police for a passport, for it is not allowed to any but an inhabitant of Paris, and whose name is registered as such, to be in the city without one. I found the place out on the bank of the river—an enormous building containing an infinity of offices; and it was only by paying for information that I found out the one I wanted. On entering it I beheld a large chamber containing about twenty clerks with enormous books before them, and a great number of people on the outside of the tables, all of whom came on business respecting passports. Mine was a peculiar case, and soon gained attention, for, excepting Sir H. Davy's, there was not another free Englishman's passport down in the books. An American, who was there and (perceiving me at a loss for French) had spoken to me, would scarcely believe his senses when he saw them make out the paper for a free Englishman, and would willingly have been mighty inquisitive. After having numbered the passport, and described me in their books with a round chin, a brown beard, a large mouth, a great nose, &c. &c., they gave me the paper and let me go.

It was a call upon all magistrates and authorities to respect, aid, &c.; but the article which pleased me most, as having a great appearance of liberality, was that, as a stranger who had not always opportunities, I was to be admitted, on showing the paper, to all public property—as museums, libraries—on any day, though the public are admitted to many of them but two or three times a week.

Sunday, 13th.—I went this morning into some of the churches, but was not induced to stop long in any of them. It could hardly be expected that they would have attractions for a tasteless heretic. Some of them were very large and finely ornamented inside, and more particularly the altars. Gold shone in abundance, and the altar-pieces or pictures were by the best masters. Masses were performing in many of them, sometimes two or more in one church at different altars, though at the same time. There were many people in some of them, but numbers seemed, like me, to be gazers. A theatrical air spread through the whole, and I found it impossible to attach a serious or important feeling to what was going on.

Tuesday, 23*rd*.—MM. Ampère, Clément, and Desormes came this morning to show Sir H. Davy a new substance, discovered, about two years ago, by M. Courtois, saltpetre manufacturer. The process by which it is obtained is not yet publicly known. It is said to be procured from a very common substance, and in considerable quantities.

A very permanent and remarkable property of this substance is, that when heated it rises in vapour of a deep violet colour. This experiment was shown by the French chemists, and also the precipitation of nitrate of silver by its solution in alcohol. Sir Humphry Davy made various experiments on it with his travelling apparatus, and from them he is inclined to consider it as a compound of chlorine and an unknown body.

It was in small scales with a shining lustre, colour deep violet, almost black ; its appearance was very like plumbago. When sublimed it condensed again, unaltered, into crystals. A very gentle heat is sufficient to volatilise a portion of it, for when the bottle containing it was held in the hand, the interior soon became of a violet colour. It dissolves very readily in alcohol, and forms a solution of a deep brown colour, which precipitates nitrate of silver, and a portion of the precipitate laid on paper in the sun's light was rapidly discoloured. When a portion of it was rubbed with zinc filings in contact with the atmosphere, a fluid combination was formed. When treated with potassium in a glass tube, they combined with inflammation. When it was heated in contact with phosphorus, a strong action took place, and an inflammable gas came over. On removing the retort from the mercurial apparatus, dense fumes issued from it, which seemed to be muriatic acid; they had the same odour, and precipitated nitrate of silver in the same way. When the iodine was placed in contact with mercury, a combination was gradually formed, which, on being heated, became first orange-coloured, then black, and at last red.

Unfortunately for me, I as yet know nothing of the language, or I should have learned much more concerning this singular substance; but thus I have marked down most of its principal characters. A future day may produce something further about it; Sir Humphry Davy now thinks it contains no chlorine.

Wednesday, 24*th*.—Being indoors all day, I amused myself by noticing in what the apartments we occupy differ from English rooms. The most striking difference in this cold weather is in the fires and fireplaces. Wood is the universal fuel.

Charcoal is the usual fuel of the kitchens, and almost the whole of the business done on the Seine is with that article. The river is divided between it and the washerwomen.

In the internal decoration of apartments the French apply glass and marble, two beautiful materials, in much greater abundance than the English do. In brass working, also, they have risen to great perfection, and their application of this material to the construction of ornamental time-pieces is exceedingly ingenious and beautiful.

French apartments are magnificent, English apartments are comfortable; French apartments are highly ornamented, English apartments are clean; French apartments are to be seen, English apartments enjoyed; and the style of each kind best suits the people of the respective countries.

Saturday, 27th.—A short search in the booksellers' shops gave me a little idea of the state of the trade in Paris. My object was a French and English grammar; but they were scarce, not owing to a want of books in general, but to a want of communication between the two nations. I at last found one composed for Americans, and that answered my purpose. Stereotype printing is in great vogue here, and they have many small books beautifully done. The French type is squarer and more distinct than the English.

Books are very cheap here in proportion to English books; I should think, on an average, they are scarcely half the price, and yet large private libraries are seldom met with. Bibliomania is a disease apparently not known in France; indeed, it is difficult to conceive how their light airy spirits could be subjected to it.

Wednesday, Dec. 1st.—On this and the preceding day Sir H. Davy made many new experiments on the substance discovered by M. Courtois. . . . .

M. Clément has lately read a paper on it at the Institute, in which he says it is procured from the ashes of sea-weeds by lixiviation and treatment with sulphuric acid: he conceives it to be a new supporter of combustion.

The discovery of this substance, in matters so common

and supposed so well known, must be a stimulus of no small force to the inquiring minds of modern chemists. It is a proof of the imperfect state of the science even in those parts considered as completely understood. It is an earnest of the plentiful reward that awaits the industrious cultivator of this the most extensive branch of experimental knowledge. It adds in an eminent degree to the beautiful facts that abound in it, and presents another wide field for the exercise of the mind. Every chemist will regard it as an addition of no small magnitude to his knowledge, and as the forerunner of a grand advance in chemistry.

Friday, 3rd.—I went to-day to the laboratory of M. Chevreul, at the Jardin des Plantes, with Sir H. Davy, where we remained some time at work on the new substance. I observed nothing particular in this laboratory, either as different to the London laboratories or as peculiarly adapted to the performance of processes or experiments. It was but a small place, and perhaps only part of the establishment appropriated to chemistry.

Wednesday, 8th.—I went to-day with Sir H. Davy to L'École Polytechnique, the national school of chemistry, to hear the *leçon* given to the scholars. It was delivered by M. Gay-Lussac to about two hundred pupils. The subject was vapour, and treated of its formation, electricity, compressibility, &c. Distillation both by heat and cold was introduced. It was illustrated by rough diagrams and experiments, and occupied about an hour. My knowledge of French is so little I could hardly make out the lecture, and without the experiments I should have been entirely at a loss.

. . . . . .

FARADAY TO HIS MOTHER.

'Thursday, December 9, 1813. Received June 4, 1814.

'Dear Mother,—I write at this time in hopes of an opportunity of shortly sending a letter to you by a person who is now here, but who expects soon to part for England. It has been impossible for me to write before since we have been in France, but you will have heard of me from Mr. Brande, and I expect also from Mrs. Farquhar. I feel very anxious to know how you are situated in your house, and the state of your health, but see no mode at present by which you can convey the desired information except by Mr. Brande. Sir Humphry told me that when Mr. B. wrote to him he would send in the same letter an account of your health, and I expect it impatiently. It would be of no use to write a long letter, as it is most probable it would not reach you. We are at present at Paris, but leave it shortly for the south of France, and Lyons will be our next resting-place. . . .

'I could say much more, but nothing of importance; and as a short letter is more likely to reach you than a long one, I will only desire to be remembered to those before mentioned, not forgetting Mr. Riebau, and tell them they must conceive all I wish to say.

'Dear Mother, I am, with all affection, your dutiful son,

'M. FARADAY.

'Mrs. M. Faraday, 18 Weymouth Street, Portland Place.'

The journal continues thus :—

Saturday, Dec. 11*th.*—Sir Humphry Davy had occasion to-day for a voltaic pile to make experiments on the new substance now called iodine, and I obtained one from M. Chevreul. . . . .

Saturday, 18*th*.—This was an important day. The emperor has just visited the senate in full state. The weather has been very bad; but that did not prevent me and thousands more from going to see the show. I went, about twelve o'clock, to the Tuileries Gardens, and took my station on the terrace, as being the best place then vacant. After waiting some time, and getting wet through, the trumpet announced the procession. Many guards and many officers of the court passed us before the emperor came up, but at last he appeared in sight. He was sitting in one corner of his carriage, covered and almost hidden from sight by an enormous robe of ermine, and his face overshadowed by a tremendous plume of feathers that descended from a velvet hat. The distance was too great to distinguish the features well, but he seemed of a dark countenance and somewhat corpulent. His carriage was very rich, and fourteen servants stood upon it in various parts. A numerous guard surrounded him. The empress and a great number of courtiers, &c., followed in other carriages. No acclamation was heard where I stood, and no comments.

Tuesday, 21*st*.—I am quite out of patience with the infamous exorbitance of these Parisians; they seem to have neither sense of honesty nor shame in their dealings. They will ask you twice the value of a thing, with as much coolness as if they were going to give it you; and when you have offered them half their demand, and, on their accepting it, you reproach them with unfair dealings, they tell you 'you can afford to pay.' It would seem that every tradesman here is a rogue, unless they have different meanings for words to what we have.

I was very much amused, for half an hour this morning, in observing the operation and business of a noted shoeblack at the corner of the passage running under the theatre (Feydeau, I think). The shop has two entrances; the interval between is well glazed, and preserved in as neat order as the windows of a coffee house. Along the back of the shop run benches covered with cushions: they are four or five feet from the ground, and a foot board runs at a convenient distance beneath them. When a customer enters he takes his exalted seat, and generally a newspaper (two or three lying constantly in the shop), and a spruce shopman immediately makes his feet look the best part about him. The place is well lighted up; and the price of all these enjoyments, for a soft seat, news, brilliant boots, &c., is 'what you please!'

Wednesday, 29*th*.—This morning we left Paris, after a residence in it of three months, and prepared ourselves for new objects and new scenes. The morning was fine, but very cold and frosty; but on entering the forest of Fontainebleau we did not regret the severity of the weather, for I do not think I ever saw a more beautiful scene than that presented to us on the road. A thick mist which had fallen during the night, and which had scarcely cleared away, had, by being frozen, dressed every visible object in a garment of wonderful airiness and delicacy. Every small twig and every blade of herbage was encrusted by a splendid coat of hoar frost, the crystals of which in most cases extended above half an inch. This circumstance, instead of causing a sameness, as might have been expected, produced an endless variety of shades and forms—openings in the foreground placed far-removed objects in view which, in their airy dress and softened by distance, appeared as clouds fixed by the hands of an enchanter; then rocks, hills, valleys, streams, and woods; then a milestone, a cottage, or human beings, came into the moving landscape, and rendered it ever new and delightful. We slept this night at Nemours.

Thursday, 30*th*.—Though cold and dark, we were on our way to Moulins by five o'clock this morning; and though somewhat more south than London, yet I do not perceive any superior character in the winter mornings here. However, as we always judge worse of a bad thing when it is present than at any other time, I may have been too cross with the cold and dark character of our early hours. The moon had set—a circumstance to be regretted, for though assisted only by the faintness of starlight, yet I am sure our road was beautiful: 'twas along the banks of the river within a few yards of the water, which indeed at times came to our horses' feet. On our left was a series of small hills and valleys lightly wooded, and varied now and then by clustering habitations. These dark hours, however, have their pleasures, and those are not slight which are furnished at such times by the memory or the imagination. I have often regretted the interruption caused by the change of horses, or the mending of broken harness. 'Tis pleasant to state almost audibly to the mind the novelty of present circumstances—that the Loire is on my right hand, that the houses to the left contain men of another country to myself, that it is French ground that I am passing over; and then to think of the distance between myself and those who alone feel an interest for me, and to enjoy the feeling of independence and superiority we at present possess over those sleeping around us. We seem tied to no spot, confined by no circumstances, at all hours, at all seasons, and in all places we move with freedom—our world seems extending and our existence enlarged; we seem to fly over the globe, rather like satellites to it than parts of it, and mentally take possession of every spot we go over.

Saturday, Jan. 8th.—Reached Montpellier to-day at a very good hour, about two o'clock. The weather has been very cold and frosty all the morning, more so I think than at any time before this winter, but the sky is beautifully clear and brilliant. We have passed many olive plantations, and are now in a country famous for fine oil. I believe we shall remain here some time, and have opportunity to notice the country. The town seems to be very pretty; but the hotel we are in must not be compared to that at Lyons, except for good oil, and wine, and good-nature in all the persons in it.

Wednesday, Feb. 2nd.—Since we have been here, Sir Humphry has continued to work very closely on iodine. He has been searching for it in several of the plants that grow in the Mediterranean, but has not obtained certain evidence of its presence. If it exists in them at all it is in very minute quantities, and it will be scarcely possible to detect it.

Sunday, 6th.—The Pope passed through this place a few days ago, on his way to Italy. He has just been set at liberty. The good Catholics have, in expectation of his coming, been talking of his sufferings and troubles for many days past, and at every hour felt their curiosity and devotion rise higher. At last he came, not to stop in the town, as was supposed, but merely to pass by the outside of it. Early in the morning the road was well peopled, and before ten o'clock almost every person in the town was there but myself. They say he was received in a very pathetic manner, and with a multitude of sighs, tears, and groans. Some people accompanied him for miles from the town, and some had in the morning gone many miles to meet him.

Thursday, 17*th*.—Left Nice this morning, and advanced towards the Alps by a road on the sides of which were gardens with oranges and lemons in great profusion. We soon entered among the mountains: they were of limestone stratified very regularly, and appearing at a distance like stairs. At some distance up we came to a place where the strata for many yards consisted of small pieces of limestone an inch in size, more or less cemented together by carbonate of lime. Varieties occurred here and there, and in these places the cement had taken a stalactite form. A dropping well added to the variety of objects, and, appearing in a very picturesque situation, added much to the beauty of the scene.

Saturday, 19*th*.—Col de Tende. Rose this morning at daybreak, which was much advanced at half-past five o'clock, and made preparations for crossing the great mountain, or Col de Tende. At Tende the noble road, which had given such facile and ready conveyance, finished, and it was necessary to prepare for another sort of travelling. Expecting it would be very cold, I added to my ordinary clothing an extra waistcoat, two pairs of stockings, and a nightcap: these, with a pair of very strong thick shoes and leathern overalls, I supposed would be sufficient to keep me warm.

About nine o'clock, horses were put to the carriage, and we proceeded towards the mountain by a road which, though not so good as the one we passed, was by no means bad, and still continued by the river of yesterday. On each side were extensive plains covered with snow to a great depth, but sufficiently hard and solid to support the men who accompanied, or rather who guided, us as they walked upon it.

There were at present but two of these persons, the chief and one of the sixty-five composing the band. They walked on before, whistling and helping; and the scene, so strange and singular to us, never attracted their attention, unless to point out to us the site of an avalanche or a dangerous place.

There was something pleasant in the face and appearance of the chief, and I thought him a good specimen of the people here. He was a tall man, not at all thick, but his flesh seemed all muscle and strength. His dress consisted of few articles—trowsers, a loose waistcoat, an open jacket, a hairy cap, very heavy-soled shoes, and coarse gaiters, or overalls, tied round his shoes to keep out the snow. This was all his clothing, and I found his comrades just like him. His gait was very peculiar, contracted, I suppose, by walking constantly on the snow, where a firm footing is required.

The road began to change soon after leaving Tende, and at last became nothing but ice. It was now fit for beasts of burden only: grooves had been formed in it at equal distances to receive the feet of the horses or mules, and prevent their falling; and though convenient to them, it was to us a great evil, for as the wheels fell successively into the ruts, it produced a motion not only disagreeable, but very dangerous to the carriage. Sir H. Davy here pointed out to me the rocks of micaceous schist, and I learned at the same time that granite is always found under this rock. The only vegetation visible, though there might be much under the snow, was of pine trees. They lifted their verdant tops above the snow, and in many places broke the monotony of a white landscape. Having passed some distance on this road, we were suddenly stopped by the wheels being entangled in the snow, which was full two-and-a-half feet high on each side of our way; and it was a work of no small labour to disengage them again. Having at last got free, we again pursued our route. The day was fine and clear, and the sun darted his burning rays with much force upon us, so as even to make us throw off our great coats; and though here encompassed by fields of snow and ice, they did not, apparently, produce any cooling effects, but seemed merely to increase the splendour of a brilliant day. Rocks here granite.

We were now joined by four or five of the gang, who had advanced to meet us and to give aid, if necessary, on the road; and in about half an hour afterwards we came to a halt, and the end of the carriage road. Here on an open space the rest of the men who were to conduct and convey us and the baggage over the mountain were collected, and the scene was a very pretty subject for the pencil. On one side lay three or four *traîneaux*, or sledges, and further on two *chaises-à-porteur*, or chairs mounted on sledges. Many men were engaged in unloading and reloading mules that had come over the mountains; and at some distance I saw a person coming down who had crossed from the other side, and who had two men to sustain him. This made me suppose that the passage was a very bad one, and, as I intended to walk to preserve some little warmth, raised my expectations in no small degree.

The horses being taken off, all hands worked to dismount the carriage and charge the *traîneaux*, and after some time this was done. The pieces of the carriage were placed on two sledges, and the rest, as the wheels, boxes, &c., loaded five mules. In this place the barometer stood at 27 inches, and the thermometer, in the shade of my body, was at 46° F.; but the instrument had been in the carriage all the morning, and was heated by the intense power of the sun in the fore part of the day.

The *traîneaux* with the body of the carriage had started about twelve o'clock. After they had been loaded, ropes were fixed to them at different parts, and they were consigned each *traîneau* to about twenty men, who were by main strength to haul it over the mountain. They set off with a run and loud huzzas; but the mules were not ready until one o'clock, and as a mule driver could be better spared, if wanted, than a man from the sledges, I kept in their company. At one o'clock we began to ascend the mountain, and I commenced walking with a barometer in my hand, the scale of which ran from 24 to 18 inches. The path quickly changed its appearance, and soon became not more than eighteen or twenty-four inches wide. Being formed by the constant tread of mules, it consisted merely of a series of alternate holes in the snow, each of which was six or eight inches deep, and ten or twelve across: in one part of our route the path had

1814. Æt. 22.

been formed on the snow on so steep an ascent that the surface exposed in a perpendicular direction was above four times as broad as the width of the path. Marks of feet were perceived crossing the mule path here and there, but leading directly up to the top of the mountain. These were the steps of the persons who had taken charge of the *chaises-à-porteur*; and the ascent must have been a very singular one to the person carried, who would often be placed in a position nearly vertical from the steepness of the ascent. In other places the marks of the carriage *traîneaux* were visible. They had passed over plains of snow undirected by any previous steps, or aught else except the devious mule path and the top of the mountain. At a distance, and nearly at the top of the mountain, the *chaises-à-porteur* were just visible, and a bird soaring below it the men pointed out to me as an eagle.

After some climbing and scrambling, the exertion of which was sufficient to keep me very comfortably warm, I reached a ruined, desolate house, half-way up the mountain. Here we found the *traîneaux*; the men, having rested themselves after this long and laborious stage, were now waiting for their leader and the dram bottle. From hence the view was very extensive and very singular. The mules, which I had left at a little distance behind me, appeared winding up the staircase, which itself, towards the bottom, seemed to diminish to a mere line, and all was enclosed in an enormous basin, and shut out from everything but the skies. The sound of the men's voices and the mule bells was singularly clear and distinct.

After a short rest, all resumed their labour; and at forty-three minutes after three o'clock I gained the summit of the mountain, having been three hours ascending. Here, at a height of more than 6,000 feet above the level of the sea, the thermometer was at 11° F., and the barometer at 25·3 inches. The observation of the barometer was made by Sir H. Davy, for though the mercury oscillated in the instrument I carried, it did not fall within the scale. The summit of the mountain is very pointed, and the descent consequently begins immediately on the other side; but I stopped a few minutes to look around me. The view from this elevation was very peculiar, and if immensity bestows grandeur was very grand. The sea in the distance stretching out apparently to infinity, the enormous snow-clad mountains, the clouds below the level of the eye, and the immense white valley before us, were objects which struck the eye more by their singularity than their beauty, and would, after two or three repetitions, raise feelings of regret rather than of pleasure. The wind was very strong and chilling, and during the short time I remained (not a quarter of an hour) we were enveloped in a cloud, which, however, soon passed off, and left all clear before us.

To descend was a task which, though not so tedious, was more dangerous than to ascend. The snow was in much greater quantity on this side the mountain than on the other, and in many places where it had drifted assumed a beautifully delicate appearance. In numberless spots it was, according to the men, more than twenty feet deep, and in descending it often received me more than half-way into it. In some parts caves or hollows occur, having only a small hole in the top of the apparently solid snow; and those who leave the mule path and descend directly down the mountain, must be particularly careful of such places, lest they fall into them and be lost. In descending, one of the mules missed the steps, and fell rolling over and over several yards down the side of the mountain; fortunately it was not hurt, and by cutting a temporary path in the snow, and supporting its burden on each side, it was quickly brought into the right road.

After I had been descending for some time down the mountain, the men with the *traîneaux* made their appearance at the top, having finished by far the most arduous part of their undertaking. They stopped only to change the arrangement of the cords, and were almost immediately in motion. Their progress now was extremely quick, and I thought dangerous, for men and *traîneaux* actually slid in a direct line towards the bottom of the mountain, over these extensive and untried plains of snow.

About half-past four we passed a little village consisting of seven or eight huts nearly buried in the snow; they were uninhabited, and are principally intended as a refuge for the men if accidents or other circumstances should occur in the mountains during the night. At about a quarter past five evening began to come on, and the effect produced by it on the landscape was very singular, for the clouds and the mountains were so blended together that it was impossible to distinguish the earth from the atmosphere. The *traîneaux* now rapidly approached us with surprising velocity; and as it began to grow dark, I joined them, there being the greatest number of men, had I wanted aid, and their hard work being finished.

Just as the starlight came on, the sounds of the evening bell of a distant village were faintly heard.

They came from the place we were going to. Lanterns were lighted, and one was carried before each *traîneau*; and guided by them and a river which owed its birth to the mountain, and was here of considerable size, we got to Leman about seven o'clock in the evening, and there put up for the night; supper and rest being both welcome.

Tuesday, 22nd.—To-day we remained in Turin. It happened to be the last of the Carnival, so I walked out in the afternoon to see what was doing in public. Towards three o'clock the shops were shut up very rapidly, and the masters betook themselves to walking, gazing, and the amusements now going on. Such as were determined to be cheerful in spite of appearances joined the number who were waltzing to the music of itinerant musicians; and certainly these did not seem the least cheerful and happy part of the population of Turin (I may perhaps add, also, were not the least numerous). I strolled to one place just on the skirts of the town, and found it crowded by those who thus easily obtained their pleasures. It was a large clear piece of ground on the bank of a branch of the Po, and resounded from end to end and side to side with the harmony of a number of musical professors. The little groups into which they had formed themselves were surrounded each one by its circle of ever-moving and never-tired dancers, and the spaces between these groups were filled up by a heterogeneous mixture of singers, leapers, boxers, chestnut merchants, applestalls, beggars, trees, and lookers-on. I fell in with one of the most worthy sets—at least, they claimed the preeminency, and it was allowed them by the other mobs. The nucleus was an enormous stone, on which stood tottering five musicians, and twenty-one pairs were waltzing round them.

Returning into the town, I found that those of Turin who were superior to the vulgar amusements I have just described had resorted to the employment which custom has ascertained to be more refined and suited to their ordinary habits and occupations. That such a suitableness exists I verily believe, but I think I perceived much more cheerfulness, and means much better suited to produce it, in the crowds I had left than in those I came to see; but pride will supply many wants, and food, clothing, amusement, and comfort are very often given up for its peculiar gratification.

I found myself in a wide and spacious street of considerable length, terminating at one end in a large place having a church in its centre. All the entrances into this street were guarded by soldiers, and no person on horseback or in a carriage could gain admission into it except at the top. A long string of carriages, curricles, saddle-horses, &c., filled it, and they continued to move on progressively up and down the street and round the church for several hours. It was presumed that these vehicles carried the principal persons of the town, but nobody pretended to say that the owners were actually in them. One of very goodly aspect and fine appearance was pointed out to me when coming up; the horses were very handsome, and the coachman and footman as spruce as could be—and so were the two maids in the inside. The next was not so dashing, but it was empty; and the third was so shabby that I did not look to see what was inside. There were, however, an immense number of persons who stood on each side of the street, looking and gazing with great apparent satisfaction, and who, if they had been conscious of the comparison I was then making between the scene before me and the one I had just left, would have looked down upon me with contempt and derision, no doubt, equal at least to that which at the same moment occupied my mind. Silly, however, as the whole affair was, it had nearly led to circumstances of more importance. A gentleman in his curricle, attended by his servant, had come down one of the side streets, and wished to enter the *corso* unlawfully, but was stopped by the soldier guarding the entrance. The gentleman, irritated by the repulse, endeavoured to force his way by rough driving. The soldier set his bayonet, and stood his ground. The horse was slightly wounded and near being killed, and from the pain became restive and had nearly killed his master, who was in the end obliged to turn back with his wounded horse, amidst the derision and laughter of the surrounding mob.

Saturday, 26*th*.—Genoa. In the evening I went to the opera; the performance was for the benefit of a principal actress, and, in consequence, an addition was made to the common course of entertainments. At a moment when the actress had completed the performance of a difficult piece of singing, and had begun to receive abundance of applause, a shower of printed papers descended from the top of the theatre amongst the audience—some of them were copies of the piece just sung, and others were verses in praise of the actress. Like the rest I strove to obtain one, and succeeded. After the shower of papers, several pigeons were thrown, one by one, from the top of the theatre into the pit, and some of them suffered cruel deaths. Before the evening concluded, a repetition of these entertainments took place, accompanied by a shower of gold (paper), with all of which the audience appeared highly delighted. The theatre was small and pretty; the performance to me very tedious.

Friday, Feb. 4th.—To-day went with Sir H. Davy to the house of a chemist to make experiments on torpedoes. There were three small ones, being about five inches long and four broad. They were very weak and feeble, for, when the water in which they were was warmed, they gave but very weak shocks, so weak that I could not feel them, but Sir H. Davy did. The great object was to ascertain whether water could be decomposed by the electrical power possessed by these animals. For this purpose wires were cemented into tubes, and the surface of the ends only exposed, as in Dr. Wollaston's method, and the two extreme ends, connected with plates of tin, were placed in contact with the two organs of the fish. It was then irritated, and often contracted, apparently giving the shock, but no effect on the water was perceived. However, the smallness and weakness of the fish and the coldness of the season prevented any negative conclusion from being formed; and it was resolved to complete the experiments in more advantageous circumstances another time.

Saturday, 5th.—The weather as yet against our voyage, and in the afternoon a storm of thunder and lightning, and rain with water-spouts. A flash of lightning illuminated the room in which I was reading, and I then went out on the terrace to observe the weather. Looking towards the sea, I saw three water-spouts, all depending from the same stratum of clouds.

I ran to the sea-shore on the outside of the harbour, hoping they would approach nearer, but that did not happen. A large and heavy stratum of dark clouds was advancing apparently across the field of view, in a westerly direction: from this stratum hung three water-spouts—one considerably to the west of me, another nearly before me, and the third eastward; they were apparently at nearly equal distances from each other. The one to the west was rapidly dissolving, and in the same direction a very heavy shower of rain was falling, but whether in the same place, or nearer or more distant, I could not tell. Rain fell violently all the time at Genoa. The one before me was more perfect and distinct in its appearance. It consisted of an extended portion of cloud, very long and narrow, which projected, from the mass above, downward, in a slightly curved direction, towards the sea. This part of the cloud was well defined, having sharp edges, and at the lower part tapering to a point. It varied its direction considerably during the time that I observed it—sometimes becoming more inclined to the horizon, and sometimes less; sometimes more curved, and at other times more direct. Beneath the projecting cloud, and in a direction opposite to the point, the sea appeared violently agitated. At the distance it was from me I could merely perceive a vast body of vapour rising in clouds from the water, and ascending to some height, but disappearing as steam would do, long before it reached the point of the cloud. The elongated part apparently extended from the stratum about $\frac{2}{5}$ths or $\frac{1}{4}$th of the distance between it and the water; but no distinct and visible connection, except in effect, could be perceived between the vapour of the sea and the extended cloud.

Appearances were exactly the same with the third water-spout. The first disappeared very quickly; the second continued, after I saw it, about ten or twelve minutes, and the third fifteen or twenty minutes. They continued their progressive motion with the cloud during the whole time; and the third, before it disappeared, had advanced considerably—I should think two or three miles. The destruction and dissolution of the water-spout seemed to proceed very rapidly when it had once commenced, and three or four minutes after the apparent commencement of decay it had entirely disappeared—the vapour, the sea, and the cloud diminishing in nearly equal proportions. They were situated much further out at sea than I at first supposed—I should think five or six miles; and, of course, what I have here noted is merely a relation of the thing as it appeared to me, and is possibly very different from the real truth. During the time I remained on the port or quay observing the water-spouts, a strong flash of lightning and a heavy peal of thunder proceeded from the same stratum of clouds. . . . .

Monday, 21*st*.—Florence. Went with Sir H. Davy to the Academy del Cimento. Saw the library, which was very small. Then saw the gardens, which are conveniently arranged; and afterwards saw the rooms and the apparatus. Here was much to excite interest —in one place was Galileo's first telescope, that with which he discovered Jupiter's satellites. It was a simple tube of wood and paper, about 3½ feet long, with a lens at each end. The field of view very small. There was also the first lens which Galileo made. It was set in a very pretty frame of brass, with an inscription in Latin on it. The lens itself is cracked across.

The great burning-glass of the Grand Duke of Tuscany, a very powerful instrument, was here also. The academy has abundance of electrical apparatus. One electrical machine was made of red velvet, the rubber being formed of gilt leather. In a glass case were preserved some Leyden phials perforated in an extraordinary manner. One of them had been broken through, by the discharge of a battery, in such a manner that a hole of a quarter of an inch in diameter had been formed. On the edge of the hole the glass was absolutely pulverised, and cracks extended to the distance of a quarter of an inch from it, beyond which nothing particular was visible. The tin foil had been burnt off a space larger than the extent of the cracks. Another jar had also been perforated, and a hole of some size formed; but the most particular thing was the combustion of the tin foil, which had taken place from the side to the bottom of the jar, and a surface of about seven square inches exposed. This was on the outside; on the inside the foil was but little damaged. There was also a numerous collection of magnets of various forms and combinations—an enormous one, which was enclosed in a box, supported a weight of 150 lbs.

Sir H. Davy afterwards went into the laboratory, to which place I attended him, and made various experiments on iodine.

Tuesday, 22*nd*.—The day principally employed in the laboratory.

Wednesday, 23*rd*.—The same as yesterday.

Thursday, 24*th*.—Prepared things to-day for the combustion of the diamond in oxygen to-morrow, if the weather prove fine. The Duke's great lens was brought out and placed in the garden, and its effect observed on wood, &c., which it instantly inflamed.

Sunday, 27*th*.—To-day we made the grand experiment of burning the diamond, and certainly the phenomena presented were extremely beautiful and interesting. A glass globe containing about 22 cubical inches was exhausted of air, and filled with very pure oxygen procured from oxymuriate of potash; the diamond was supported in the centre of this globe by a rod of platinum, to the top of which a cradle or cup was fixed, pierced full of holes to allow a free circulation of the gas about the diamond. The Duke's burning-glass was the instrument used to apply heat to the diamond. It consists of two double convex lenses, distant from each other about 3½ feet; the large lens is about 14 or 15 inches in diameter, the smaller one about 3 inches in diameter. The instrument is fixed in the centre of a round table, and is so arranged to admit of elevation or depression, or any adjustment required, at pleasure. By means of the second lens the focus is very much reduced, and the heat, when the sun shines brightly, rendered very intense. The instrument was placed in an upper room of the museum; and having arranged it at the window, the diamond was placed in the focus, and anxiously watched. The heat was thus continued at intervals for three quarters of an hour (it being necessary to cool the globe at times), and during that time it was thought that the diamond was slowly diminishing and becoming opaque. Now we had only a partial spectrum, for the upper part of the window obstructed the sun's rays; but having sunk the whole of the apparatus, it was again exposed, and a very strong heat obtained. On a sudden Sir H. Davy observed the diamond to burn visibly, and when removed from the focus it was found to be in a state of active and rapid combustion. The diamond glowed brilliantly with a scarlet light inclining to purple, and when placed in the dark continued to burn for about four minutes. After cooling the glass, heat was again applied to the diamond, and it burnt again, though not nearly so long as before. This was repeated twice more, and soon after the diamond became all consumed. This phenomenon of actual and vivid combustion, which has never been observed before, was attributed by Sir H. Davy to the free access of air. It became more dull as carbonic acid gas formed, and did not last so long. The globe and contents were put by for future examination.

Monday, 28*th*.—To-day we endeavoured to repeat the experiment of yesterday, but the sun had sunk too low, and sufficient heat could not be obtained. The experiment was then made, substituting chlorine for oxygen, but no change of the diamond was produced; the platina was slightly acted on, and a cork used to support the prop was very much corroded, but no combination of carbon and chlorine was effected.

Tuesday, 29*th*.—This morning, the diamond which we had ineffectually endeavoured to burn yesterday was brought out, and, being again exposed in the focus as before, produced by a bright and powerful sun, soon exhibited signs of strong electricity, and a few small filaments which floated in the gas were variously attracted and repelled by it. In a short time a thin slip of platina, used to fasten the diamond in, was observed to fuse, and in taking the globe from the focus the diamond was seen in a state of intense ignition and combustion. It continued burning for some time, and when extinguished was again ignited by the application of the lens, and burnt as before. This was repeated five times, after which no visible signs of combustion or diminution could be perceived. Having applied the heat for some time longer, it was at length taken away, and the temperature reduced to the same point as at the commencement of the experiment; but no vapour nor any signs of the formation of water could be perceived. A portion of the gas was then put into a curved tube and decomposed by potassium; charcoal was liberated, and no other substance but the charcoal and the oxygen could be found in the gas formed. Limewater was then introduced, and the carbonate of lime precipitated carefully gathered, washed, and preserved for further analysis and investigation.

Wednesday, 30*th*.—A diamond was to-day exposed in the focus of the instrument in an atmosphere of carbonic acid, to ascertain whether carbonic oxide would be thus formed; but after exposure to the heat from twelve to one o'clock no change could be perceived, nor had the diamond lessened in weight.

Saturday, April 2*nd*.—Another diamond was burnt in oxygen gas, for the purpose of ascertaining whether any azote or other gas inabsorbable by water was given off; but all the circumstances are against such a supposition. The apparatus was very simple and convenient. A small globe with a neck was filled entirely with water, and then $\frac{2}{3}$rds with oxygen gas; the diamond, fixed on the end of its support, was introduced, and the globe transferred by means of a wineglass to the lens. When heated, the diamond burned as in the former cases, giving out a fine scarlet light inclining to purple, and extreme heat. The globe was then allowed to cool to its former temperature; but no apparent change of volume had occurred, though the water must have absorbed a little carbonic acid gas (the volume was not noticed with precision). A solution of lime, and afterwards of caustic potash, was then introduced into the globe, and the carbonic acid removed; then, by adding nitrous gas to the same volume of the remaining gas and the oxygen used, a portion of which had been preserved, it was found that the diminution was exactly the same, and the remainder very small in both cases; so that nothing but carbonic acid gas had been formed, and as yet it appears that the diamond is pure carbon.

Having finished these experiments, we bade adieu for a time to the Academy del Cimento, and prepared to depart for Rome.

Tuesday, 7*th.*—Rome. Went into a bookseller's shop to inquire for an Italian and English dictionary, but could not find one. Went into the workshop of a bookbinder, and saw there the upper part of a fine Corinthian pillar of white marble, which he had transformed into a beating stone of great beauty. Found my former profession carried on here with very little skill, neither strength nor elegance being attained.

FARADAY TO HIS MOTHER.

'Rome: April 14, 1814.

'My dear Mother,—It is with singular pleasure I commence writing after so long a silence, and the pleasure is greatly increased by the almost certainty that you will get my letter. We are at present in a land of friends, and where every means is used to render the communication with England open and unobstructed. Nevertheless, this letter will not come by the ordinary route, but by a high favour Sir H. Davy will put it with his own, and it will be conveyed by a particular person.

'I trust that you are well in health and spirits, and that all things have gone right since I left you. . . . . Mr. Riebau and fifty other friends would be inquired after could I but have an answer. You must consider this letter as a kind of general one, addressed to that knot of friends who are twined round my heart; and I trust that you will let them all know that, though distant, I do not forget them, and that it is not from want of regard that I do not write to each singly, but from want of convenience and propriety; indeed, it appears to me that there is more danger of my being forgot than of my forgetting. The first and last thing in my mind is England, home, and friends. It is the point to which my thoughts still ultimately tend, and the goal to which, looking over intermediate things, my eyes are still directed. But, on the contrary, in London you are all together, your circle being little or nothing diminished by my absence; the small void which was formed on my departure would soon be worn out, and, pleased and happy with one another, you will seldom think of me. Such are sometimes my thoughts, but such do not rest with me; an innate feeling tells me that I shall not be forgot, and that I still possess the hearts and love of my mother, my brother, my sisters, and my friends. When Sir H. Davy first had the goodness to ask me whether I would go with him, I mentally said, "No; I have a mother, I have relations here." And I almost wished that I had been insulated and alone in London; but now I am glad that I have left some behind me on whom I can think, and whose actions and occupations I can picture in my mind. Whenever a vacant hour occurs, I employ it by thinking on those at home. Whenever present circumstances are disagreeable, I amuse myself by thinking on those at home. In short, when sick, when cold, when tired, the thoughts of those at home are a warm and refreshing balm to my heart. Let those who think such thoughts useless, vain, and paltry, think so still; I envy them not their more refined and more estranged feelings: let them look about the world unincumbered by such ties and heart-strings, and let them laugh at those who, guided more by nature, cherish such feelings. For me, I still will cherish them, in opposition to the dictates of modern refinement, as the first and greatest sweetness in the life of man.

'I have said nothing as yet to you, dear mother, about our past journey, which has been as pleasant and agreeable (a few things excepted, in reality nothing) as it was possible to be. Sir H. Davy's high name at Paris gave us free admission into all parts of the French dominions, and our passports were granted with the utmost readiness. We first went to Paris, and stopped there two months; afterwards we passed, in a southerly direction, through France to Montpellier, on the borders of the Mediterranean. From thence we went to Nice, stopping a day or two at Aix in our way; and from Nice we crossed the Alps to Turin, in Piedmont. From Turin we proceeded to Genoa, which place we left afterwards in an open boat, and proceeded by sea towards Lerici. This place we reached after a very disagreeable passage, and not without apprehensions of being overset by the way. As there was nothing there very enticing, we continued our route to Florence; and, after a stay of three weeks or a month, left that fine city, and in four days arrived here at Rome. Being now in the midst of things curious and interesting, something arises every day which calls for attention and observations. The relics of ancient Roman magnificence, the grandeur of the churches, and their richness also—the difference of habits and customs, each in turn engages the mind and keeps it continually employed. Florence, too, was not destitute of its attractions for me, and in the Academy del Cimento and the museum attached to it is contained an inexhaustible fund of entertainment and improvement; indeed, during the whole journey, new and instructive things have been continually presented to me. Tell B. I have crossed the Alps and the Apennines; I have been at the Jardin des Plantes; at the museum arranged by Buffon; at the Louvre, among the *chefs-d'œuvre* of sculpture and the masterpieces of painting; at the Luxembourg palace, amongst Rubens' works; that I have seen a GLOWWORM!!! water-spouts, torpedo, the museum at the Academy del Cimento, as well as St. Peter's, and some of the antiquities here, and a vast variety of things far too numerous to enumerate.

'At present I am in very good health, and so far is travelling from disagreeing with me that I am become somewhat heavier and thicker than when I left England. I should have written to you long ago, but I had no hopes of getting a letter conveyed; but at present I conclude that you will surely have this. I have a thousand things more to say, but do not know how to select one from the other, so shall defer them all to a more convenient opportunity. When you write into the country, remember me, if you please, to all friends there, and more particularly to those to whom I have written. At present, I bid farewell for a time to all friends, wishing them much happiness.

'I am, dear Mother, with earnest wishes for your health and welfare, your dutiful son,

'M. FARADAY.

'PS. There is no certain road open at present by which you can write to me, so that, much as I wish it, it must be deferred a little longer. We have heard this morning that Paris was taken by the Allied troops on March 31, and, as things are, we may soon hope for peace, but at present all things are uncertain. Englishmen are here respected almost to adoration, and I proudly own myself as belonging to that nation which holds so high a place in the scale of European Powers.

'Adieu, dear Mother, at present. Your dutiful son,

'M. FARADAY.'

Sunday, April 19th.—Went to-day to the palace of the Monte Cavallo, situated on the summit of Mount Quirinal; it is the most pleasantly situated of any in Rome, and gives a fine view of the city. The interior is interesting at this moment for its unfinished state. It was intended for the palace of the future King of Rome, son of Napoleon I. The apartments in which the works of ornament and luxury have been commenced are numerous, and the designs are extremely appropriate and beautiful. Ancient columns of beautiful materials have been cut up to form the fireplaces, and the entrances, and the ceiling of each chamber contains a beautiful painting. The design in the ceiling of the chamber intended for the emperor represented night.

A bard was sleeping over his harp, and dreams of war, conquest, and glory hovering around him. The idea and the execution were beautiful, but the application not very felicitous. Mosaic work both in wood and stone abounded, and no expense has, indeed, been spared in the commencement of these works. The place was shown us where the Pope was taken prisoner, and the door and staircase by which the soldiers entered.

From this place we went to the Villa Borghese, and passed a pleasant hour in the gardens, which abound with fountains, temples, and statues.

I hardly know why I endeavour to describe the antiquities and works of art that I see in the course of my walks; I know very well that I can give no idea of their beauty or value, and that the observations I make would appear absurd to others. But this journal is not intended to instruct and inform, or to convey an imperfect idea of what it speaks of; its sole use is to recall to my mind at some future time the things I see now; and the most effectual way to do that will be, I conceive, to write down, be they good or bad, or however imperfect, *my present impressions.*

Thursday, May 5th.—Went to-day to repeat an experiment first made by Signor Morrichini of Rome, on magnetism, and interesting in the highest degree from its novelty, and the important conclusions it leads to.

The experiment consists in giving magnetism to a needle by the solar rays, and was thus performed :—a needle was fixed on the point of a pin of brass by a piece of wax, in a direction north and south nearly, and a spectrum being formed by the decomposition of a strong ray of white light, the violet rays were collected by a lens, and the focus gradually drawn along the needle, beginning at the middle and proceeding to the north point. A white screen was placed behind the needle, which rendered it easy to bring the focus where it was wanted. The focus was thus made to pass continually over the needle, always in the same direction, for an hour; and then the sun getting too high for our apparatus, the experiment was finished for this day. On essaying the needle by iron filings, and also by suspension, no effect was perceived. This was attributed to the misty air, for the sun was not so bright as it very often is here at this season. The result of the experiment when made successfully—that is, with a bright sun—is a magnetic needle, which points north and south, attracts iron filings, attracts the contrary pole of a common magnet, and repels the same pole, and possesses in every respect the same qualities. The experiment has been repeated above fifty times, and always with success—sometimes being completed in half an hour, and sometimes requiring the bright sun of two days. It is found that only the violet, the blue, and the green rays have this power—the violet most, and the green least. The red rays have been thrown in the same way on a needle at various times for twenty-four hours, but they produce no effect. The experiment also succeeds as well in winter as in summer, and in general is finished sooner. I saw one needle which had been thus formed. It was highly magnetic, attracted strongly iron, and quickly took its direction when suspended, observing the variation as other needles do. The needle which was taken for this experiment was reserved for another day, when the process would be continued.

1814.
ÆT. 22.

Saturday, 7th.—Rose at twelve o'clock last night, and at two this morning was on the road to Naples. This early hour was chosen for the purpose of proceeding as far as possible on the first day, for, as the road is considered as very dangerous and abounding with robbers, it was necessary to take some precautions in order to ensure our arrival at Naples. At the second change of horses, six gendarmes joined us, and escorted us over a dangerous part of the road. At the next post the number was lessened, but some were with us all day. Ruins were plentiful in this day's journey, and two fine aqueducts appeared. The Colosseum, at the commencement of the journey, was beautiful in the extreme, and, as the moon, nearly full, appeared through the upper range of arches, had a romantic and beautiful appearance.

Friday, 13th.—Mount Vesuvius was the employment of to-day, and fully rewarded the trouble and fatigue attendant upon seeing it. We were at the foot of the mountain by half-past eleven o'clock. From hence it is usual to proceed to what the peasants call the foot of the summit on asses, but I walked this road. The lower part of the mountain is very highly cultivated, and yields grapes, figs, and other fruit in abundance. This luxuriant vegetation is continued upwards to a considerable height, and takes place upon a soil of lava, partly decomposed, and partly pulverised. The road is very disagreeable from the quantity of large loose stones. After crossing an ancient stream of lava, we came at length to the Hermitage, or half-way house. The recluse came out to meet us, but though in a black gown, he proved himself not at all deficient in the art of an innkeeper. We stopped there a short time, enjoying the extensive view of both sea and earth presented to us, and then continued our route upwards, until we had reached the foot of the summit. This last road was very rough and hilly, laying over streams of lava, which in many places appeared broken, or thrown together in a very singular manner. At this place the most tiresome part of our journey commenced. What they call the summit—i.e. the mountain formed by the ashes thrown out, and which contains the crater—is constituted by lava and dust. The streams of lava that issue forth at each eruption partly cool on the summit, and remain there, and are afterwards covered by the ashes and stones thrown out. This collection of course has the altitude naturally taken by a heap of small rolling bodies, and added to this great degree of inclination, it has the disadvantage of being a very bad foundation for the feet, continually receding as the foot advances; nevertheless, by the aid of strong sticks, and two or three restings, we attained the top by about half-past two o'clock. Here the volume of smoke and flame appeared immense, and the scene was fearfully grand. The ground beneath us was very hot, and smoke and vapour issued out from various spots around us. On the top of the summit rises a small mountain, which from a distance appears covered with sulphur. This we ascended, and then came to a resting-place, from whence the mouth of the volcano, and part of the crater were visible. From here we had a fine view of the fire. The wind was very favourable, and blew the smoke from us, and at times we could see the flames breaking out from a large orifice with extraordinary force, and the smoke and vapour ascending in enormous clouds; and when silence was made the roaring of the flames came fearfully over the ear. We then advanced to a piece of ground thrown up on the edge of the crater, and were then within 100 (feet?) of the orifice from whence the flames issued forth. Here we had a fine view of the crater, appearing as an enormous funnel, and the smoke issuing forth in abundance from most parts of it. It was incrusted in many places with the same yellow substance before observed, and which Sir H. Davy said was muriate of iron. After having stood here a few minutes, we were obliged to retreat with rapidity, for the wind, changing suddenly, brought the smoke upon us, and the sulphurous acid gas threatened suffocation. I incautiously remained to collect some of the substances, and was then obliged to run over the lava, to the great danger of my legs. Having gained our former station, we remained there for a time to observe things in more security.

There appeared to me to be two very distinct species of smoke or vapour—that which proceeded from the mouth of the volcano was very dense, of a yellow-white colour, and rolled away in the form of cumuli. From the odour of that which had been thrown on us by the wind, it appeared to consist principally of sulphurous acid gas and water. From other places a white vapour arose which disappeared rapidly as steam would do, and from the faint odour it possessed appeared to be very little else but steam. Sir H. kindly explained to me that all, or nearly all, the water which was condensed by the mountain, and which would otherwise form streams and springs, was volatilised by the heat, and was one principal cause of the smoke. On the spot where we were a very considerable heat was evident, and in cavities in the lava it was too strong for the hand to bear; and a boy who came up with us cooked some eggs by this heat, and laid them out with bread and wine as a repast. In these cavities a very evident odour of muriatic acid and chlorine was perceptible, and the various substances of white, red, and yellow colours appeared to be muriate of iron.

Where the heat was not too great, the ashes at the top were very moist, from the condensation of water volatilised from below. At one spot where we were, a man poured some wine into a hole, where the heat was so great as to cause a strong ebullition, and the wine immediately evaporated. At some little distance from this spot a white sublimate appeared on the lava in certain spots which proved to be muriate of soda. Having observed everything that was visible at this time, we began, at about three o'clock, to descend, and found the task as disagreeable almost as the ascent, though much more rapid. The asses appeared at an immense distance below us, and the space between them and us was a steep plain of rolling ashes. The descent of this part reminded me very strongly of the descent from the Alps, and the principal difference was, that in the last case we sank and rolled in snow, and in the present in ashes.

Having, however, continued to slide to the bottom, we again got on a less inclined path, and proceeded towards the Hermitage. In the descent the streams of lava, which at various times had issued out from the mountain, appeared before us as rivers, and were extremely distinct from the nude soil by their black and barren appearance. The lava appeared exactly as I expected it would do. In many places the liquid or soft portion beneath had been covered by loose masses, which occasioned a rough and rugged surface; and in some parts it had flowed clear and uncovered, and had taken the various curved forms and marks where it had met an obstacle. It was of various colours and densities, graduating from the densest kind to almost pumice.

About the middle of the mountain were various fragments of a green primitive rock, which Sir H. said had been shot out of the mountain. In about two hours we gained the bottom, and proceeded homewards.

Saturday, 14*th*.—To-day was again devoted to Vesuvius; but the party was much larger, and the hour much later than yesterday, the intention being to see it in the night. We were at the foot of the mountain by half-past four o'clock, and gained the summit by about half-past seven o'clock. During our climb upwards many beautiful views were highly enjoyed, and the evening light on the mountains and promontories was very fine. Some rain fell as we approached the top, which being volatilised by the heat made a much greater appearance of vapour than yesterday; added to which the fire was certainly stronger, and the smoke emitted in far greater quantity. The wind had changed since yesterday, but was still very favourable to our intention, and carried the smoke and vapour in a long black line over the hills to a great distance. It now became dark very quickly, and the flames appeared more and more awful—at one time enclosed in the smoke, and everything hid from our eyes; and then the flames flashing upwards and lighting through the cloud, till by a turn of the wind the orifice was cleared, and the dreadful place appeared uncovered and in all its horrors. The flames then issued forth in whirlwinds, and rose many yards above the mouth of the volcano. The flames were of a light red colour, and at one time, when I had the most favourable view of the mouth, appeared to issue from an orifice about three yards, or rather more, over.

Cloths were now laid on the smoking lava, and bread, chickens, turkey, cheese, wine, water, and eggs roasted on the mountain, brought forth, and a species of dinner taken at this place. Torches were now lighted, and the whole had a singular appearance; and the surrounding lazzaroni assisted not a little in adding to the picturesque effect of the scene. After having eaten and drunk, Old England was toasted, and 'God save the King' and 'Rule, Britannia' sung; and then two very entertaining Russian songs by a gentleman, a native of that country, the music of which was peculiar and very touching.

Preparations were now made for the descent; so taking an earnest view of the crater, we began, at half-past eight, to slide down as before, but with an increase of difficulties, for the uncertain and insufficient light of a waving and fickle torch was not enough to show rightly the path. And the increase of number caused an increase of evils, for, not proceeding in a line, those before ran great danger from the rapid descent of large fragments loosened by those behind; and the cries of alarm were very frequent—and, indeed, some that I saw would have endangered the life of any person against whom they might have struck. Having, however, reached the bottom of the summit, the asses were mounted, and at about eleven o'clock we found ourselves in the village at the foot of the mountain. During our descent, the beautiful appearance of the fire frequently drew the attention of all persons, and the long black cloud, barely visible by the starlight, appeared as a road in the heavens.

Got home by half-past eleven o'clock, highly pleased and satisfied with the excursion.

Friday, June 3rd.—Remained at Terni to-day, employing the time in an excursion to the waterfall of the same name. The cascade is situated at the distance of five miles from the village, and is considered the finest and highest in Europe. It is formed by the lake Velino, which falls into the Nera from a height of above 200 feet. Ascending from the village, we passed first by the river after its fall, which was much swelled by the rains, and increased my expectations of the cascade. Its colour was very white and turbid. After a while we arrived at the summit of the fall, and saw the stream, of great size, pass with impetuosity to the edge of the fall. Turning a little on one side, we came to a spot where the fall was visible: and here truly the scene was beautiful. The view of the country alone is very fine, and from a great eminence a fine view of the valley and the distant mountains is obtained. But the fall is what attracts the first notice, and calls the attention with an immense roaring. The rocks are perpendicular, and the water falls nearly free in a stream of the purest white. The force with which it descends causes a considerable quantity to be dispersed in the air in mists and fine rain; and this produced the beautiful phenomena of the rainbow in the utmost perfection. Our situation was the most advantageous for this effect, and the prismatic colours appeared with extreme vividness in an arc of nearly two-thirds of a circle of about 200 feet in diameter. The red outer and more faint rainbow was also visible. The water in its descent carried a vast body of air with it into the recess below, which was forced up again by the curling stream. This was produced in a very curious and singular manner. The water thrown up from the bottom of the fall by the concussion condensed and gathered together in small streams, which ran down the sides of some low rocks opposite to the fall, and situated in the stream; but these streams, on passing over the projecting parts of the rock into the free air, were arrested by the ascending current of air, and were broken into minute drops, and disappeared. Advancing still further, we came to a small summer-house, or arbour, at the end of a projecting rock, which gave us a direct and opposite view of the fall, which here appeared in full force and grandeur. After admiring the scene for a while, we walked through a beautiful country to the lake from whence the stream proceeded. The vegetation here is extremely luxuriant, and woodbine, geraniums, myrtles, thyme, mint, peppermint, &c., scented the air in the walk. The nature of the old lake was shown in a very interesting manner. The base of this part is travertine or calcareous matter deposited by water, which appeared in strata and as stalactites: in many places agates appeared in the limestone. They had surrounded a nodule of limestone as a thick shell, and afterwards had been inclosed by a further deposition of calcareous matter. Chert also occurred. This cavity was formerly the bottom of a lake which deposited these matters; but a passage was cut for the waters, and by this means a great extent of country recovered. We came at length to the present lake, and taking a boat rowed on it for some time to enjoy the scenery. The water of the river is slightly opaque, but the lake is beautifully clear. It is surrounded by mountains of fine form and situation, and the views are delicious. In the distance appeared the mountains which separate the kingdom of Naples from the Pope's dominions, and snow was observed on them in considerable quantities. On returning to that part of the lake from which we set out, we observed fishermen dragging. Here many water plants were growing beneath the surface of the water; and I observed many streams of oxygen gas ascend from them, liberated by the rays of a bright sun. We now returned to the fall, and again enjoyed our former views; but the rainbow was still more beautiful than before, for, from the respective situation of the sun, the eminence, and the fall, a bow of three-fourths of a circle was attained. We descended the mountain to gain the bottom of the fall, and in going there passed through a grove of orange trees in full blossom, and which scented the gale to a distance of many yards. And at a spot nearly opposite to the bottom of the fall, we found that it was not one fall only, but three successive falls that conducted the water to the bottom, the first only being visible from the summit. The other falls are not so high as the first, and take place amongst enormous rocks and masses, breaking the water into two streams, which unite again at the bottom. The scene here was fine, but not so beautiful as above.

We now left the fall, and went to the side of the river, to observe the deposition which is constantly going on here. The masses of travertine were enormous, forming ledges over the present streams, and appearing in various singular forms. The different extraneous bodies which fall into the stream are soon incrusted by the water, and in some places masses of leaves are found which are thus covered, and give their form to stone. Some poles which had been placed in the water to form a ledge were incrusted nearly to the thickness of half an inch. The waters here are very white. Having enjoyed this interesting place for some hours, we returned to the village on the hill, and from thence to Terni.

. . . . . .

Sunday, 5th.—Continued our way through a fine but hilly country, well cultivated and very gay in appearance. We left the Apennines about midday, and entered the valleys, where the scenery was beautiful from an accompanying river and the surrounding mountains. Towards the evening we came to a mark of the Romans, and one worthy of their name. It was the Flaminian Road, which in this place, near the mountain of Asdrubal, is cut in the side of a solid rock for the length of more than a mile and a half. Here the scenery was sublime to the highest degree, and almost terrific—the enormous overhanging cliffs rendering still darker the shades of evening, and a roaring torrent rolling at our feet. At last the way appeared entirely shut up, and we seemed as if on the point of entering the Shades as we passed into the rock; but the passage was short, and we soon gained again the open air. This astonishing outlet was cut by the Romans through the rock in order to pass a jutting point, and gives the last and highest character to this wonderful road. It is in length about twenty-five yards, and sufficiently large to admit a carriage without the least danger. The whole road has been repaired by modern hands, made larger, more commodious, and is now the means of the easiest conveyance where conveyance seemed impossible. The rocks are limestone.

On leaving this pass, we entered a country the character of which could scarcely be perceived from the advancement of the evening; but entertainment and delight were not wanting, for the fireflies appeared before and about us in innumerable quantities, and at a distance they covered the sides of the mountain, and near us they passed over the fields, hovered on the edge or crossed the road, often attaching themselves to the harness, and emitting their bright and harmless flashes of light in a rapid and beautiful manner.

Lightning of the finest kind also appeared before us on the horizon, and the evening was filled by the phenomena of light.

Friday, 17*th*.—Milan. Saw M. Volta, who came to Sir H. Davy, an hale elderly man, bearing the red ribbon, and very free in conversation.

FARADAY TO HIS MOTHER.

'Geneva: July 1. Received July 18.

'I hope, dear Mother, that you are in good health, and that nothing occurs to disturb you or render you uncomfortable, and that no changes of a disagreeable nature have happened since I left you. I hope, too, for the health and welfare of all my friends . . . . and that at some time I shall be happy with you all again. I think often and often of you, and in thoughts often enjoy your company. I contrast the company of my friends with the presence of strangers, and I compare the convenience and cleanliness of home to the want and filth of foreign accommodations. Things run irregularly in the great world; and London is now I suppose full of feasting and joy, and honoured by the presence of the greatest personages in Europe. I find reason everywhere to feel proud of my country, and find everyone ready to praise her and to honour her virtues. My thoughts run hastily, dear Mother, from one thing to another; but you must excuse it at this moment, and attribute it to the urgency of circumstances. I long for the moment when I shall salute all my friends personally; but till the moment arrives, I must be indebted to the good and kind offices of others, and now of you, dear Mother. Remember me to . . . . tenderly and affectionately; and remind Mr. Riebau, &c., that I still exist; and if Robert will call at the Institution and tell Mrs. Greenwood I wish to be remembered to her, I should feel the favour.

'Adieu, dear Mother. At present my moments are expired; but I still remain, and ever shall do on this earth, your affectionate and dutiful son,

'M. FARADAY.'

TO MR. R. G. ABBOTT.

'Geneva: Saturday, August 6, 1814. Received August 18.

'Dear Robert,— . . . . I feel too grateful for the goodness of Mr. De la Roche, of whom my mother will give you some account. If he should still be in King Street or in London, I should like to have my name mentioned to him, with thanks on my side; but he is perhaps in France, and if I see Paris again I shall search for him.

'During the time I have passed from home, many sources of information have been opened to me, and many new views have arisen of men, manners, and things, both moral and philosophical. The constant presence of

Sir Humphry Davy is a mine inexhaustible of knowledge and improvement; and the various and free conversation of the inhabitants of those countries through which I have passed has continually afforded entertainment and instruction. On entering France, the dissimilarity between the inhabitants and the people of my own country was strong and impressive, and entered firmly into my mind. I have found the French people in general a communicative, brisk, intelligent, and attentive set of people; but their attentions were to gain money, and for their intelligence they expect to be paid. Politeness is the general character of the people, a character which they well deserve; but the upper classes have carried it beyond the bounds of reason, and in politeness they lose truth and sincerity. Their manners are very insinuating and kind, their address at once easy and free, and their conversation vivid and uninterrupted, but though it dwells for hours on the same subject, you can scarcely make out what the subject is, for it is certainly the most confined, the most uninteresting and inapplicable conversation I have met with. The French language in the mouth of the people has a softness of such delicacy as is not to be found in other languages, and in which, I think, it is very superior to the Italian—this last, indeed, appears to me to have an effeminate character; but the French language has a great degree of strength and expression, and is yet delicate and tender. The Italian language is by far the easiest to learn, from the circumstance of every letter being pronounced in the words the same as when single, and also because there is a greater similarity between the words of the Italian and English languages than between those of the French and English languages. But, however, I must not pretend to judge as yet of the character of these languages with precision, for I am but little acquainted with them, though, at the same time, I have endeavoured to avoid the imputation of idleness.

'Civilisation seems to have taken different paths in the nations of Europe towards the end of, or rather latter part of, her progress. At Paris civilisation has been employed mostly in the improvement and perfection of luxuries, and oftentimes, in the pursuit, has neglected the means of adding to domestic and private comfort, and has even at times run counter to it. In ornaments, indeed, the Parisians excel, and also in their art of applying them; but in the elegance of appearance utility is often lost, and English articles which have been formed under the direction of a less refined but more useful judgment are often eagerly preferred. At Paris everything yields to appearance, the result of what is called fine taste: the tradesman neglects his business to gain time to make appearance; the poor gentleman starves his inside to make his outside look well; the jeweller fashions his gold into trinkets for show and ornament; and so far does this love of appearance extend, that many starve in a garret all the week to go well dressed to the opera on Sunday evening. I, who am an Englishman, and have been bred up with English habits, of course prefer English civilisation to the civilisation of France, and think that my common sense has made the best choice; but every-day experience teaches me that others do not think so; yet, though I have no right to suppose that I excel all those who differ from me, I still am allowed the liberty of forming my own opinion. The civilisation of Italy seems to have hastened with backward steps in latter years, and at present there is found there only a degenerate, idle people, making no efforts to support the glory that their ancestors left them, but allowing it and their works to fall into obscurity. Cramped by ignorance and buried in dirt, they seem to have been placed in a happy soil only to show forth their degeneracy and fallen state; and Rome is at this day not only a memento of decayed majesty in the ruins of its ancient monuments and architecture, but also in the degeneracy of the people.

. . . . . .

'Believe me truly and sincerely yours, with all wishes,

'M. FARADAY.'

'Geneva: Friday, August 19. Received September 12.

'Dear Mother,—It is with the greatest pleasure that I embrace every opportunity that offers the best chance of communicating with you. At this time a gentleman, a friend of Sir Humphry Davy's, leaves this place, I believe, early to-morrow, and expects to be in England in about twelve days; and shortly after the expiration of that time, I hope you will be reading over this sheet of paper. I have written many letters to you from various places, as Paris, Sestri, Genoa, Florence, Rome, &c., and also from here; but I have only received one from England, and it only notices the arrival of one of these. Most of these letters, it is true, were sent by private hands, and may yet gain you; but some have left Geneva by post, and I am much disappointed in not having yet received answers, although I attend daily at the office. The letter I received, and which came a welcome messenger from a distant beloved country, was written by Robert Abbott; nor am I deficient in gratitude for his kindness, not only as it was a proof of his own friendship and remembrance, but as it also quieted my anxieties with respect to you, though in a manner so short that it only excited stronger desires for the letter it mentioned as being written by R. (his brother). I should have written to R. long ago, but I wait for his communication. B. has also a long communication, and if it is not on the road, beg of him to despatch it immediately, if you receive this before September 1 or 2, but otherwise he must direct it to the Poste Restante, Rome.

'Here, dear Mother, all goes on well. I am in perfect health, and almost contented, except with my ignorance, which becomes more visible to me every day, though I endeavour as much as possible to remedy it. The knowledge that you have let your house, and that it has been doing its office to you almost since the day that I left you, was very pleasing; and I hope sincerely that you enjoy health and strength of mind to govern it with your accustomed industry and good order. The general assurance of A. that all friends were in perfect health was much to know; but I want a more particular detail, for, amongst so many, it is almost impossible but that some varieties and changes must occur. I must beg of you to return my thanks to ——; Mr Riebau also might not be displeased to hear the name of his former apprentice.

'It is needless, dear mother, to tell you that I wish you well, and happy and prospering—you must know that my wishes cannot be otherwise; and it is the same thing with ——; and yet, though it is needless, I cannot help but say so. I expect that we shall leave this place, where there is very little indeed, except fine weather and a beautiful view of Mont Blanc, to detain us, about the middle of September, when we shall ascend a little northward and see a little of Germany, passing round the Alps to Venice; and having seen that place for a day or two, we shall then take the most convenient road down Italy to Rome. It is Sir Humphry's intention to be at Florence about the middle of October, and at Rome about the middle of November; but till we arrive there we shall be constantly moving about, and I shall therefore be able to receive no letters, after I leave this place, until I get there: but there I shall expect to find a whole packet. As letters are about a fortnight on the road between here and London, any letters sent after the first or second of September are likely to reach Geneva after I have left it, and will then probably be lost, or very much delayed, and, consequently, you will be so good as to act accordingly. When you write into the country, remember me to all friends there, and also to all who may ask after me at home. There are some persons to whom I should be glad to be remembered, but as it is possible that such remembrances might raise unjust ideas and observations, I will delay them until I return home again. At present, dear Mother, good-bye (for when writing I seem to talk to you, and on leaving my paper it appears to me as a farewell). Farewell then, dear Mother, for a short time, when I hope again to find myself amongst you.

'Yours with the firmest affection,

'M. FARADAY.'

FARADAY TO ABBOTT.

'Geneva: September 6, 1814.

'Dear B——, It is with extreme pleasure that I pursue a correspondence which I find is not to be impaired either by time, absence, or distance—a correspondence which has been dear to me from the first moment of its existence, which I have found full of pleasure, and which I have never regretted; and its continuance continually gives me fresh proofs that it will ever remain, as it has been, a strong and irreproachable source of instruction and amusement. I thank you, dear B., as earnestly as I can do for your long and kind letter, which I shall endeavour to answer as well as I can, though not in such a manner as it ought to be. I have not, I can truly assure you, enough time to write you a letter as long as your own; I have a great deal of occupation, which leaves me but little to myself, and my journal is much behindhand; and as we leave this place in eight or nine days, I shall have difficulty in arranging my things and clearing up my papers. My head at this moment is full of thought respecting you and me—respecting your uneasy situation and mine, which is not at all times pleasant and what I expected. Your last letter has partly collected these thoughts, and I shall probably state some of them on this sheet of paper.

. . . .

'Some doubts have been expressed to me lately with respect to the continuance of the Royal Institution; Mr. Newman can probably give a guess at the issue of them. I have three boxes of books, &c. there, and I should be sorry if they were lost by the turning up of unforeseen circumstances; but I hope all will end well (you will not read this out aloud). Remember me to all friends, if you please. And "now for you and I to ourselves."

'I was much hurt in mind to hear of your ill-health, and still more so to understand your uncomfortable situation; for, from what I have felt at times, I can judge of your feelings under such a painful bondage. I am as yet but young, B., very unacquainted with the world, with men and manners, and too conscious of my ignorance to set up for a moraliser; but yet, dear friend, I have not passed on to this day without a little experience; and though not endued with the acutest powers of mind, I have been forced to notice many things which are of service to me, and may be useful to you. If they are, I shall not repent the trouble I give you; and if they are not, you must attribute them to the warmth of my feeling for you.

'You are, you inform me, in a situation where gain only is the object; where every sentiment is opposed to yours; where avarice has shut out every manly feeling; where liberal thoughts and opinions are unknown; where knowledge, except as it is subservient to the basest and lowest of feelings, is shut out; where your thoughts, if not looking to the acquisition of money, are censured; and where liberality and generosity never enter. These are things which I know to be so opposite to your mind and inclinations that I can well conceive your feelings; and, as if it were to increase those feelings, this disagreeable situation follows one that was perfectly pleasant and agreeable.

'In passing through life, my dear friend, everyone must expect to receive lessons both in the school of prosperity and in that of adversity; and, taken in a general sense, these schools do not only include riches and poverty, but everything that may cause the happiness and pleasure of man, and every feeling that may give him pain. I have been in at the door of both those schools; nor am I so far on the right hand at present that I do not get hurt by the thorns on my left. With respect to myself, I have always perceived (when, after a time, I saw things more clearly) that those things which at first appeared as misfortunes or evils ultimately were actually benefits, and productive of much good in the future progress of things. Sometimes I compared them to storms and tempests, which cause a temporary disarrangement to produce permanent good; sometimes they appeared to me like roads—stony, uneven, hilly, and uncomfortable, it is true—but the only roads to a good beyond them; and sometimes I said they were clouds which intervened between me and the sun of prosperity, but which I found were refreshing, reserving to me that tone and vigour of mind which prosperity alone would enervate and ultimately destroy. I have observed that, in the progress of things, circumstances have so worked together, without my knowing how or in what way, that an end has appeared which I would never have fancied, and which circumstances ultimately showed could never have been attained by any plans of mine. I have found also that those circumstances which I have earnestly wished for, and which ultimately I have obtained, were productive of effects very different to those I had assigned to them, and were oftentimes more unsatisfactory than even a disappointment would have been. I have experienced, too, that pleasures are not the same when attained as when sought after; and from these things I have concluded that we generally err in our opinions of happiness and misery. I condole with you, dear B., most sincerely on the uneasiness of your situation, but at the same time I advise you to remember that it is an opportunity of improvement that must not be lost in regret and repining. It is necessary for man to learn how to conduct himself properly in every situation, for the more knowledge he has of this kind the more able is he to cope with those he is at times sure to meet with. You have under your eye a copy of thousands, and you have the best opportunities of studying it. In noticing its errors you will learn to avoid them: what it has good, will by contrast appear more strongly; you will see the influence of the passions one on another, and may observe how a good feeling may be utterly destroyed by the predominance of an opposite one. You will perceive the gradual increase of the predominant sentiment, and the mode in which it surrounds the heart, utterly debarring the access of opposite feelings. At the same time, dear friend, you will learn to bear uneasy situations with more patience. You will look to the end, which may reward you for your patience, and you will naturally gain a tone of mind which will enable you to meet with more propriety both the prosperity and adversity of your future fortune. Remember that, in leaving your present situation, you may find a worse one, and that, though a prospect is fair, you know not what it may produce. You talk of travelling, and I own the word is seducing, but travelling does not secure you from uneasy circumstances. I by no means intend to deter you from it; for though I should like to find you at home when I come home, and though I know how much the loss would be felt by our friends, yet I am aware that the fund of knowledge and of entertainment opened would be almost infinite. But I shall set down a few of my own thoughts and feelings, &c. in the same circumstances. In the first place then, my dear B., I fancy that when I set my foot in England I shall never take it out again; for I find the prospect so different from what it at first appeared to be, that I am certain, if I could have foreseen the things that have passed, I should never have left London. In the second place, enticing as travelling is—and I appreciate fully its advantages and pleasures—I have several times been more than half decided to return hastily home; but second thoughts have still induced me to try what the future may produce, and now I am only retained by the wish of improvement. I have learned just enough to perceive my ignorance, and, ashamed of my defects in everything, I wish to seize the opportunity of remedying them. The little knowledge I have gained in languages makes me wish to know more of them, and the little I have seen of men and manners is just enough to make me desirous of seeing more; added to which, the glorious opportunity I enjoy of improving in the knowledge of chemistry and the sciences continually determines me to finish this voyage with Sir Humphry Davy. But if I wish to enjoy those advantages, I have to sacrifice much; and though those sacrifices are such as an humble man would not feel, yet I cannot quietly make them. Travelling, too, I find, is almost inconsistent with religion (I mean modern travelling), and I am yet so old-fashioned as to remember strongly (I hope perfectly) my youthful education; and upon the whole, *malgré* the advantages of travelling, it is not impossible but that

1814. ÆT. 22.

you may see me at your door when you expect a letter.

'You will perceive, dear B., that I do not wish you hastily to leave your present situation, because I think that a hasty change will only make things worse. You will naturally compare your situation with others you see around you, and by this comparison your own will appear more sad, whilst the others seem brighter than in truth they are; for, like the two poles of a battery the ideas of each will become exalted by approaching them. But I leave you, dear friend, to act in this case as your judgment may direct, hoping always for the best. I fear that my train of thoughts has been too dull in this letter; but I have not yet attained to the power of equalising them, and making them flow in a regular stream. If you find them sad, remember that it was in thinking on you they fell, and then excuse them.

. . . . . .

'Sir Humphry works often on iodine, and has lately been making experiments on the prismatic spectrum at M. Pictet's. They are not yet perfected, but from the use of very delicate air thermometers, it appears that the rays producing most heat are certainly out of the spectrum and beyond the red rays. Our time has been employed lately in fishing and shooting; and many a quail has been killed in the plains of Geneva, and many a trout and grayling have been pulled out of the Rhone.

. . . . . .

'I need not say, dear Ben, how perfectly I am yours,

'M. FARADAY.'

Thursday, Oct. 13th.—Vicenza. . . . . Begging has increased wonderfully since we left Germany; indeed, it is almost the birthright of modern Italy. All are beggars in some way—the innkeepers by the postilion and by the ostlers beg, and the poor people of the country universally lay aside their work and run by the carriage begging. A shepherd will leave his flock at half a mile distant to beg at the road-side as the carriage passes, and the women will leave their huts and occupation to beg at the door. The children who are more agile and brisk, commence by certain ceremonies. Some will, on seeing a carriage at a distance, lie down in the middle of the road and kiss the ground, then rising on their knees, they remain in a praying position until it comes up to them; and on its arrival, they run to the side, and beg in monosyllables of Carità, caro Dio, &c., &c. Unless sent away with something more to their liking than words, they will follow the carriage for nearly a mile. Others dexterously tumble head over heels five or six times, till the carriage approaches, and then proceed as the former ones; and others vary the ceremonies still further. This detestable habit of universal begging is imbibed by the children even with the mother's milk; and if they are not ready and prompt in their supplication, they are punished by the parent and enjoined to still further efforts. It must produce a humiliating and depressing effect on the mind of the people in general, and appears as a curse spread over the country.

Thursday, Oct. 20th.—Pietra Mala. On arriving this morning we made a halt, to see the remarkable phenomenon called in this country Il fuoco della pietra mala, or Fuoco di ligno, which is about a mile to the south of this little village. The account that I had heard of this phenomenon was as follows,—namely, that from a certain spot of ground, about ten or twelve feet square, situated on the side of the mountain, and which appeared to be the same with the neighbouring soil, several flames of various sizes broke forth and were constantly burning. Some of these flames were about a foot in diameter and height, and others were not more than an inch or two in measure. In some places they were blue, in others red; and in the nights of rainy weather so bright as to illuminate the neighbourhood to a considerable distance. If water was thrown on the flame, it would extinguish it for an instant, but soon it would spring up with redoubled force,—that combustible bodies could be easily burnt by the flame, but that the soil there appeared the same as in the neighbourhood, and, when not actually covered by the flame, not at all hot or even warm,—that M. Bernouille says water will easily extinguish it, but that the country people say, heavy rains only serve to augment its force, and that it was considered as the remains of an ancient volcano.

Also that there were two other places in the neighbourhood of Pietra Mala, but in different directions, where flames were observed at times, but rarely; and that at another place a fountain existed the waters of which were continually boiling, and inflamed on putting a light to them. This fountain was called L'Acqua bollente.

Though it was raining hard, yet that of course would not deter Sir Humphry from visiting these places; but, at the same time, it made us wish to be as quick as possible. Sir Humphry therefore went to the first place, and I went to the Acqua bollente, conducted by a man of the village, who carried some fire, some straw, and some water. I found the place in a cultivated field, not far from a mountain apparently of limestone. It was simply a puddle, perhaps formed by the present showers of rain. Much gas rose from the earth, and passed through the water, which made it appear boiling, and had given rise to its name; but the water and the ground were quite cold. I made another puddle with the water we brought, near the one I found there, and I saw that the gas rose up through it also; and it appeared to be continually passing off from a surface of more than eighteen inches in diameter. The soil appeared deep, and close to the spot supported vegetation readily. The man inflamed some straw and then laid it on the ground; immediately the gas inflamed, and the flame spread to some distance from the straw over the surface of the earth, waving about like the flame of weak spirits of wine: this flame burnt some moments. On putting a light to the bubbles which rose through the water they inflamed, and sometimes a flame ran quickly from them over the whole surface of the water. I filled a bottle with the gas, but I could not distinguish any smell in it. In pouring water into the bottle and lighting the jet of gas that came out, a large clear flame was obtained. The whole of this flame was a very pale blue, like spirits of wine. It inflamed paper and matches readily, as might be expected; and when I held a dry bottle or knife over it, they appeared to become dim by condensing water: but this was uncertain, as the weather was so rainy. The water had no taste, and seemed pure rain water. I brought some of it and the gas away, and returned to the village.

Sir Humphry Davy told me that it was exactly similar to the one he had been to, except in the size of the flame, which was at his place near four feet in height and diameter. The men extinguished it three or four times, and it did not take fire spontaneously. It inflamed immediately on applying a light, and burnt with a blue colour similar to spirits of wine. It rose from the earth between some stones; but Sir Humphry saw no volcanic remains in the neighbourhood. The gas and the water were preserved for experiments at Florence.

Wednesday, 26th.—To-day a few hours were spent in the examination of the gas from Pietra Mala. The experiments were made at home, but the imperfection of the instruments admitted of no accurate examination. It was detonated in a long closed tube with oxygen over mercury by means of phosphorus, and it appeared probable that it contained carbon and hydrogen; but no certain results were obtained.

Thursday, 27th.—In the now almost deserted laboratory of the Florentine Academy, Sir Humphry to-day made decisive experiments on the gas. The detonating tube was made with platina wires, inserted in it to take the spark, and various detonations were performed. $2\frac{3}{4}$ths of the gas, which appeared to have remained unaltered from the time I had collected it, were detonated with $5\frac{1}{4}$th of oxygen, and diminished to 3; and by agitating the remaining gas with a solution of pure potassa, it diminished to half a part. It appeared, therefore, that water and carbonic acid had been formed, and Sir Humphry concluded from the proportion that the gas was light hydrocarburet, pure. When detonated with $2\frac{1}{2}$ times its volume of chlorine, it diminished to about 1, and charcoal was deposited on the sides of the tube.

Saturday, 29th.—We left Florence this morning,

about seven. Women who ride here—and there are great numbers that do constantly—sit across the horses. We passed three genteelly-dressed ladies to-day who were riding so. They seemed good horsewomen: one of them had a very restless pony to manage. . . .

Monday, November 7th.—This morning a man was punished by the civil authority close to our house, and the mode deserves notice from its singularity and its cruelty. I observed, the first day we came here, a gallows fixed up in the Corso next to our palace, which was not up when we were here before. It was about thirty or thirty-six feet high; a large pulley was fixed at the end of it, and a strong rope ran through the pulley. A stone was fixed in the earth behind the gallows with an iron ring in it. This morning the crowd began early to collect about the place, and I learnt that some one was to suffer; and about nine or half-past, the man was brought by a guard to the spot. He was placed under the gallows with his hands tied behind him, and the rope which belonged to the instrument was fixed to his hands. He was then drawn up to the top and let down again, and was thus mounted three times. It generally happens that from the way in which the man's hands and arms are placed—namely, behind him—dislocation of the shoulders ensues, and the man is crippled for life; and this effect is more certainly produced as the crime of the man is greater. In the present case the man was drawn gently up, and let as gently down again; but sometimes the motion is so rapid and violent as to put the sufferer to the greatest agony, and twist his whole body with pain. I saw the man after the punishment was finished. He was carried out of a neighbouring house where some restorative had been given to him. He was surrounded by an armed guard. His hands were tied behind, and on his breast was hung a large tablet with *Per insolenza al militare* on it. I heard that he had thrown mud at some soldiers. He seemed but little hurt by the punishment, and walked away with a firm, steady, quick pace.

Saturday, 12*th*.—It is positively affirmed, that the civil punishment above spoken of has given so much offence to the people in general that the Pope has ordered it to be abolished.

FARADAY TO HIS MOTHER.

'Rome: November 10, 1814. Received January 17.

'Dear Mother,—Time goes very strangely with me—sometimes it goes quick, and at other times the same period seems to have passed slowly; sometimes it appears but a few days since I wrote to you, and the next hour it seems like months and years. This is owing to the nature of the mind of man, which, looking to what is at a distance when occupied by present circumstances, sees it not in its true form and state, but tinged by its own cast and situation. But though thus volatile and apparently unstable, yet I am at all times, dear Mother, glad when I can by any means make an opportunity of writing to you; for though however short the distance of time since the last letter may be, yet I have always something I should like to say; and, indeed, the moment a letter is out of my hands I remember something forgotten. It is now a long time, if I may trust to my feelings and the mode of measuring time, since I had any communication with you except by thought, and indeed longer than I expected, for I was in hopes of finding at the post-office here at least three or four letters for me; but as they are not come, I content myself with anticipating the pleasure yet to be enjoyed of perusing them. Since I wrote to you or to England, we have moved over a large and very interesting space of ground. On leaving Geneva, we entered Switzerland, and traversed that mountainous and extraordinary country with health and fine weather, and were much diverted with the curious dresses and customs of the country. When I come home (unless M.'s knowledge in geography, &c. anticipate me), I shall be able to amuse you with a description, but, at present, time (excuse the excuse) will not allow me.

'From Switzerland we passed through the States of Baden, on the lake of Constance (they are very small), across an arm of the kingdom of Wurtemburg, and into Bavaria. In this route we had seen, though slightly, Lausanne, Vevay, Berne, Zurich, Schaffhausen, and the falls of the Rhine in Switzerland and Munich, and many other towns in Germany. On leaving Munich, we proceeded to and across the Tyrol, and got to Padua, and from Padua to Venice. . . . After seeing Venice for three days we left it, and came towards Italy, passing Bologna and Florence.

'I am always in health, generally contented, and often happy; but, as is usual in every state of life, wish for that I have not, but most for my return home. I envy you the pleasure you must enjoy in each other's conversation, and from which I am excluded; but I hope you will ameliorate this deprivation as far as you can by thinking at home of me. I mean quickly to write to — —, but, in the meantime, I should be happy to express through you my feelings to them:

1814. ÆT. 23.

they cannot for a moment doubt me, but at all times the testimony of remembrance is grateful and pleasing. When you see — —, give them my love in the most earnest manner you can; though, indeed, it is scarcely necessary, for they, and —, and yourself, dear Mother, must be conscious that you constantly have it when anything, as a letter, &c., reminds you and them of me; but paper is now scarce, and time advanced, and I must quickly leave this letter that it may come to you. But again, dear Mother, I beg of you to let me know quickly how you are, and how situated, as soon as possible. If right, present my humble respects to Mr. Dyer, and my remembrances to Mr. Riebau and other friends, not forgetting — —.

'Adieu, dear Mother, for a short time. As ever, your dutiful affectionate son,

'M. FARADAY.'

FARADAY TO ABBOTT.

'Rome: Saturday, November 26, 1814.

'What have I done or what have I said that I am to hear no more from England? Day passes after day, week after week, but passes without bringing me the long-wished-for letters. Did you but know the pleasure they give me, did you but know the importance they are of to me, certain I am that compassion would induce you to write. Alone in a foreign country, amongst strangers, without friends, without acquaintances; surrounded by those who have no congenial feeling with me, whose dispositions are opposite to mine, and whose employments offend me—where can I look for pleasure but to the remembrances of my friends? At home I have left those who are dear to me from a long acquaintance, a congeniality of mind, a reciprocal feeling of friendship, affection, and respect, as well for their honour as their virtues; here I find myself in the midst of a crowd of people, who delight in deceiving; are ignorant, faithless, frivolous, and at second sight would be my friends. Their want of honour irritates me, their servility disgusts me, and their impertinence offends me; and it is with a painful sensation I *think* of my friends when I remember I cannot do more. Why, then, do you delay so long that which is the greatest service you can do me? And since I have lost your company, let me at least have your thoughts; since I cannot see you, let me see the work of your hand. Through my own imprudence I have lost for a time that source I did possess, for I have left at Geneva, with books, those letters I have received from you, my brother, and yours, and which I ought never to have separated so far from me. It is possible I may never see them again, and my fears tell me I may never receive any more, and even that the possibility exists of my being for ever separated from England. Alas! how foolish perhaps was I to leave home, to leave those whom I loved, and who loved me, for a time uncertain in its length, but certainly long, and which may perhaps stretch out into eternity! And what are the boasted advantages to be gained? Knowledge. Yes, knowledge; but what knowledge? Knowledge of the world, of men, of manners, of books, and of languages—things in themselves valuable above all praise, but which every day shows me prostituted to the basest purposes. Alas! how degrading it is to be learned when it places us on a level with rogues and scoundrels! How disgusting, when it serves but to show us the artifices and deceits of all around! How can it be compared with the virtue and integrity of those who, taught by nature alone, pass through life contented, happy, their honour unsullied, their minds uncontaminated, their thoughts virtuous—ever striving to do good, shunning evil, and doing to others as they would be done by? Were I by this long probation to acquire some of this vaunted knowledge, in what should I be wiser? Knowledge of the world opens the eyes to the deceit and corruption of mankind; of men, serves but to show the human mind debased by the vilest passions; of manners, points out the exterior corruptions which naturally result from the interior; of books, the most innocent, occasions disgust, when it is considered that even that has been debased by the corruptions of many; and (knowledge) of languages serves but to show us in a still wider view what the knowledge of men and of manners teaches us.

'What a result is obtained from our knowledge, and how much must the virtuous human mind be humiliated in considering its own powers, when at the same time they give him such a despicable view of his fellow-creatures! Ah, B., I am not sure that I have acted wisely in leaving a pure and certain enjoyment for such a pursuit. But enough of it; I will turn to more pleasant recollections. I am so confident in you and the few friends I have in England, that I am quite sure it is not from any change of feeling, but from unfavourable circumstances, that I have not yet received any letters from you at Rome.

'I feel much interested in the Institution, and should much like to know its probable issue. . . . I hope that if any change should occur in Albemarle Street, Mr. Newman would not forget my books: I prize them now more than ever. Give my love again to your family and to mine. Adieu, dear Friend.

'M. FARADAY.'

FARADAY TO HIS ELDER SISTER, ÆT. 27.

'Rome, December 21, 1814.

'Dear Sister,—

. . . . . .

'*Saturday, December 24th.*—Hail to the season! May it bring every blessing down upon you; may it fill your hearts with gladness and your minds with contentment; may it come smiling as the morn which ushers in the glorious light of a summer's day, and may it never return to see you in sorrow and trouble! My heart expands to the idea that Christmas is come, for I know that my friends, in the midst of their pleasures, will think of me. Amongst you, Friendship will celebrate it—here 'tis Religion. You will have sincerity amongst you, and we hypocrisy. This is a season in which modern Rome shows forth her spirit; her churches (in number innumerable) are filled with the crowd, who in the same hour fill the streets with licentiousness and riot. For the last week no balls have been allowed, and no theatres or places of amusement were or are yet opened. To-night the religious rites begin at Santa Maria Maggiore (a beautiful church) in honour of the Virgin; and the child Jesus will be represented in a beautiful cradle richly decked with jewels and gold. Masses will be performed at this church all night; and to-morrow all the other churches will be open—St. Peter's amongst the rest. After to-morrow, the Pope loses his power for a week or more, and the Carnival begins; and this Carnival raises all my expectations, for the accounts I have heard of it make it a scene of confusion and folly. Professed fools (deserving of the title) parade the streets, and hold fearful combats armed with sugar plums. Religious clowns and every other kind of character fill the streets, and the whole world goes in masks. The theatres are opened. Puppet-shows shine in every corner, and the Italian character blazes in its full vigour. Such are the scenes with which I am surrounded; but I draw from them to contemplate those I fancy passing at home, in which I hope to join again, and which to me will recur with tenfold pleasure.

'But, dear Sister, though this frivolous spirit occupies the whole mind of a modern Roman, and debases that empire which once stood like a Colossus over the whole world, yet still this city, the seat of that empire, draws forth involuntary awe and respect.

'How often I have wished that Mr. D. could see what I saw, that he could wander with me over the mighty wilderness of ruins the Colosseum presents—sometimes mounting, sometimes descending; walking in the steps of the ancient Romans, and leaning against the walls which resounded with their voices. Again, the ancient baths, each rich as palaces and large as towns: here their paintings are to be seen in their original station, the marble which they had worked, and the walls which they had formed. Again, the columns of Titus and Antoninus, or rather of Marcus Aurelius, enormous in size, covered with beautiful sculptures, and formed of marble.

'Again, a thousand other objects, as tombs, temples,

statues, pyramids, pillars, roads, &c., which continually fill the eye of a stranger. D. would be delighted with them, and his art and skill would enable him to bring faithful ideas of them home.

. . . . . .

'God bless the little one, and you all together. I shall never feel quite happy till I get amongst you again. I have a thousand things to say, but I do not know which to say first; and if I followed my mind I should never get to an end.

. . . . . .

'Adieu, dear Sister, for a time; and believe me to be, ever and unalterably, your loving, and I hope beloved, brother, 'M. FARADAY.'

FARADAY TO HIS YOUNGER SISTER, ÆT. 12.

'Rome, December 29, 1814.

'Dear Margaret,—I am very happy to hear that you got my letter, and I am as happy to say that I have received yours. I had the last yesterday, and to-day I write you an answer. I am greatly obliged to you for the information you give me, and for the kind interest you take in my health and welfare. Give my love with a kiss to mother the first thing you do on reading the letter, and tell her how much I think on her and you. I received a letter from Mr. A. late, who told me in it that you had spent a day in his house, and he thought that you were very well pleased with it; and when you go to Mr. B.'s again, you must return humble thanks for me, and say how much I am honoured by his remembrance. I hope that all your friends are well, and I suppose that your correspondence is now very important. I am glad to

1814.
ÆT. 23.

hear that my niece E. is in favour with you, but you quite forgot to give me any news of S. I suppose thoughts of the first had put the last out of your head, which head, I fancy, has gained with these little relations a great deal of importance.

'I am also pleased to hear that you go to school, and I hope that you have enough to do there. Your writing is not improved quite so much in one year as I expected it would be when I left home; but, however, it is pretty well. Your I's are most in fault. You must make them thus, ℐ. ℐ. ℐ. ℐ. with smaller heads. My questions about Rome and Naples I did not expect you could answer, but I wished you to look into some book at school, or at Mr. Riebau's or elsewhere, and give me the answers from them, at the same time fixing them in your memory. I gave them to you as lessons, and I still hope you will learn them. I hope that you do not neglect your ciphering and figures; they are almost as necessary as writing, and ought to be learned even in preference to French. Of this last you say nothing, but I suppose you still work at it. I will tell you my way of learning the words. When in my grammar or in other books I meet with a word (and that happens often enough) that I do not know, I first write it down on a fair sheet of paper, and then look in my dictionary for its meaning; and having found it, I put it down also, but on another part of the same sheet. This I do with every word I do not know very well, and my sheet of paper becomes a list of them, mixed and mingled together in the greatest confusion—English with French, and one word with another. This is generally a morning's work. In the evening I take my list of words and my dictionary, and beginning at the top, I go regularly down to the bottom. On reading the words I endeavour to learn their pronunciation, and if I cannot remember the meaning in the other language, I look in the dictionary, and having found it, endeavour to fix it in my memory, and then go to the next word. I thus go over the list repeatedly, and on coming to a word which I have by previous readings learned, I draw a line over it; and thus my list grows little every evening, and increases in the morning, and I continually learn new and the most useful words from it. If you learn French and pursue this plan at home, you will improve in it very quickly. I must now, dear M., put an end to my letter. I have written to R., lately, and shall write to him again soon, tell him. I wish him every happiness. Give my warmest love, with your own, to mother (and say I wrote about a month ago, by favour, to her), and to R. and B. and Mr. G., and the little ones, and all your friends. Write again, at an opportunity, to your affectionate brother,

'M. FARADAY.'

Wednesday, 11*th*.—I have done nothing to-day but search for books which I cannot find—an employment which, though not successful, yet pleased me, as it took me into booksellers' shops.

Tuesday, 24*th*.—The Carnival has been the constant subject of conversation for several weeks, not only of strangers, but of the Romans themselves. Willing to give importance to their city and its diversions, they tell us what it had been and what we might expect to see; and from what they said, and from their evident anxiety for its arrival and their preparations for it, I confess I expected a great deal. 'Tis a season in which the poorest beggar will enjoy himself, even though he strip his hole of everything it contains ; and when the whole population of a city like Rome joins together for the same end, one may be allowed to expect that the end will be well attained, and especially when it is pleasure.

To-day is the first day of the Carnival, which is commenced by the long-talked-of races. I felt very anxious to see these races, for the singularity of the place, added to my having heard of them in England, raised expectation very high. But, as is generally the case with high expectations, mine were disappointed. About four o'clock I walked to the Corso, and found it very promising. It was lined on both sides with an immense number of people, some sitting on the scaffolds, and some standing about. Guards were placed on each side, at the edge of the foot pavement (or that which represented it), and the middle of the street was occupied by carriages and people that were passing up and down it (being the common Corso of Italy). Seeing that there was yet plenty of time, I walked up to the Piazza del Popolo, where the Corso begins, and from whence the horses were to start. There I paid my five baiocchi, and took my seat on a scaffold fixed under the obelisk, and which commanded a view of the whole length of the Corso ; and I amused myself for a short time in observing the preparations, which were certainly worthy of a much better thing. . . . The Corso had been lengthened by two rows of booths into the middle of the place, and was terminated by the scaffold on which I stood. Just before the scaffold several strong posts were fixed up on each side of the ground, and a very thick rope stretched across at breast high. The booths at each side were filled with persons of consequence, and the whole place was covered with carriages, the owners of which wished to peep, but could not, unless they chose to herd with the multitude. At a quarter-past four two *pint pots* were fired off, and on a repetition of the signal in about five minutes, the carriages all turned out of the Corso by the nearest side streets, and the pedestrians only remained in it. The guards now took their stations in a more orderly way; a troop of horse rode gaily to the end of the Corso, but soon returned to the commencement at full speed, and then no person was allowed to be in the open space between the guards.

The horses, five in number, were now brought out by the master and his men, who were gaudily draped on this occasion; and the horses themselves were not undecorated, though to them, as it often proves to us, their finery afterwards proved a pain.

A very slight harness made of cord, &c. was put on them, to which were attached four tin plates, one on each flank, and one on each shoulder. Over these were fixed four balls of lead, set with six or eight sharp spikes each: they hung by a string or chain five or six inches long. Their heads were decorated with a plume of feathers of various colours.

There was not much time to be spared when the horses came out; as soon as they saw the Corso and the people they were eager to set off. Six or eight men held each horse by his shoulders, his tail, his mane, &c. But with one horse these were not sufficient. He got over the starting rope and dragged the men with him, and the master cried out he could hold the horse no longer. The trumpet sounded, the rope dropped, and the animals were instantly at full speed. They took the middle of the Corso, and proceeded very directly to the other end. The plates of tin soon flew off, and the spiked balls beating the sides of the animals, and the cries of the people as they passed, were enough to frighten any English horse. They started very regularly, but one was soon six or eight yards before the others, and got in first. The mode of stopping them is by a cloth stretched across the end of the Corso, at sight of which they generally stop of themselves; but lest they should run against it and bring it down, a second is fixed up a few yards behind the first.

They profess to make the Corso open for any person's horse, but they are generally the property of one man, and trained up to the sport. It is said that, some few years ago, an English horse ran with the others, but not knowing the customs of the country, he passed the barriers and ran out of Rome. In the time of the French government the prize was 300 francs for the first horse; but the thing which repays the master is the subscription of the people in general. He parades the street for two or three days before the races, with his horses mounted by riders carrying flags, &c., and he gets abundantly sufficient to remunerate him and to leave a pretty surplus besides.

Sunday, 29th.—A mask ball takes place to-night, or, more correctly, it is to begin at one o'clock to-morrow morning, and end at six o'clock.

FARADAY TO ABBOTT.

'Rome, January 25, 1815.

'Dear B——, I begin this letter in a very cheerful state of mind, which enables me to see things with as correct an eye as it is possible for my weak judgment to do, unless, indeed, I see them too favourably; but, at all events, I hope you will not have occasion, in your answer to this letter, to repeat what, you have said in your last. I have received both the letters which you have directed to me at Rome. I have too much to say at present to waste words in thanking you for them. You know how great their value is to me, and the return I can give that will be most welcome to you is to answer them. It happens fortunately indeed that the first is in part answered, and I am not sure that I can say much more in return for it on this sheet of paper, or even on another, if I happen to extend my blotting. It was my intention when I read it to give you some account of the various waterfalls I had seen, but now I have more important and fresher subjects to treat of, and shall reserve them for another time. By important I do not mean important in itself, but only with respect to the waterfalls, and you must understand the word in that sense. I cannot, however, refrain from saying how much I feel obliged to you for your information respecting the health of my mother and our family, and hope that you will always have the charity to continue such information as far as lies in your power.

'Though it may appear somewhat consequential that I begin the letter with my own affairs, yet such is my intention at present. You found me in the last squabbling almost with all the world, and crying out against things which truly in themselves are excellent, and which indeed form the only distinction between man and beasts. I scarce know now what I said in that letter (for I have not time to take copies of them, as you supposed), but I know I wrote it in a ruffled state of mind, which, by the bye, resulted from a mere trifle. Your thoughts on knowledge, which you gave me in return, are certainly much more correct than mine; that is to say, more correct than those I sent you; which indeed are not such as I before and since have adopted. But I did not mean to give them to you as any settled opinion. They ran from my pen as they were formed at that moment, when the little passions of anger and resentment had hooded my eyes.

'You tell me I am not happy, and you wish to share in my difficulties. I have nothing important to tell you, or you should have known it long ago; but since your friendship makes you feel for me, I will trouble you with my trifling affairs. The various passions and prejudices of mankind influence, in a greater or less degree, every judgment that men make, and cause them to swerve more or less from the fine love of rectitude and truth into the wide plain of error.

'Errors thus generated exert their influence in producing still greater deviations, until at last, in many points, truth is overthrown by falsehood, and delusive opinions hold the places of just maxims and the dictates of nature. Nothing shows this truth more plainly than the erroneous estimation men make of the things, the circumstances, and the situations of this world. Happiness is supposed to exist in that which cannot possibly give it. Pleasures are sought for where they are not to be found; perfection is looked for in the place from which it is most distant, and things truly valuable are thrown aside because their owner cannot estimate them. Many repine at a situation, others at a name, and a vast multitude because they have neither the one nor the other.

'I fancy I have cause to grumble, and yet I can scarcely tell why. If I approve of the system of etiquette and valuation formed by the world, I can make a thousand complaints; but perhaps if I acted, influenced by the pure and unsullied dictates of common sense, I should have nothing to complain of, and therefore all I can do is to give you the circumstances.

'When Sir Humphry Davy first made proposals to me to accompany him in the voyage, he told me that I should be occupied in assisting him in his experiments, in taking care of the apparatus, and of his papers and books, and in writing, and other things of this kind; and I, conceiving that such employment, with the opportunities that travelling would present, would tend greatly to instruct me in what I desired to know and in things useful in life, consented to go. Had this arrangement held, our party would have consisted of Sir Humphry and Lady Davy, the lady's maid, Le Fontaine (Sir H.'s valet), and myself; but a few days before we came off, Le Fontaine, diverted from his intention by the tears of his wife, refused to go, and thus a new arrangement was necessary. When Sir H. informed me of this circumstance, he expressed his sorrow at it, and said he had not time to find another to suit him (for Le Fontaine was from Flanders, and spoke a little Italian as well as French), but that if I would put up with a few things on the road, until he got to

Paris, doing those things which could not be trusted to strangers or waiters, and which Le Fontaine would have done, he would there get a servant, which would leave me at liberty to fill my proper station and that alone. I felt unwilling to proceed on this plan; but considering the advantages I should lose, and the short time I should be thus embarrassed, I agreed. At Paris he could find no servant to suit him, for he wished for one that spoke English, French, and a little German (I speaking no French at that time), and as all the English there (ourselves excepted) were prisoners, and none of the French servants talked English, our want remained unsupplied; but to ease me he took a lacquais de place, and living in an hotel, I had few things to do out of my agreement. It will be useless to relate our progress in the voyage as it relates to this affair more particularly. A thousand reasons which I have now forgot caused the permanent addition of a servant to our family to be deferred from time to time, and we are at present the same number as at first. Sir Humphry has at all times endeavoured to keep me from the performance of those things which did not form a part of my duty, and which might be disagreeable; and whenever we have been fixed, I have had one or more servants placed under me. We have at present, although in an hotel, two men servants; but as it is always necessary to hold a degree of subordination in a human family, and as a confidential servant is also necessary to the master, and, again, as I am the person in whom Sir Humphry trusts, it obliges me to take a more active share in this part of my present occupation than I wish to do; and in having to see after the expenses of the family, I have to see also after the servants, the table, and the accommodations.

1815.
ÆT. 23.

'I should have but little to complain of were I travelling with Sir Humphry alone, or were Lady Davy like him; but her temper makes it oftentimes go wrong with me, with herself, and with Sir H.

. . . . . .

'Finally, Sir H. has no valet except myself; but having been in an humbler station, and not being corrupted by high life, he has very little occasion for a servant of that kind, and 'tis the name more than the thing which hurts. I enjoy my original employment in its full extent, and find few pleasures greater than doing so. Thus, dear B., I have answered your kind inquiries by a relation of my circumstances; things that were not of consequence enough to put in a letter before you asked for them. As things stand now, I may perhaps finish the voyage in my present company, though, with my present information, I should not hesitate to leave them in any part of the world, for I now know I could get home as well without them as with them. At all events, when I return home, I fancy I shall return to my old profession of bookseller, for books still continue to please me more than anything else. I shall now, dear friend, turn the subject, or rather change it for Philosophy, and hope, in so doing, to give you pleasure in this letter. I say this more confidently because I intend to give you an account of a paper just finished by Sir Humphry, of which one copy has already been sent by post, as a letter to the Royal Society, and all the experiments and demonstration of which I have witnessed.

'When we were at Naples the Queen gave Sir H. a pot of colour which was dug up in their presence. It contained a blue paint in powder. At Milan a gentleman had some conversation with Sir H., and gave him some pieces of blue glass from Adrian's villa at Rome; and since we have been here this time, the opportunity afforded, and the former hints, have induced Sir H. to undertake an examination of the ancient Grecian and Roman colours, with an intent to identify them, and to imitate such as were known. I shall give you a very brief account of this paper, putting down results, discoveries, and such parts as I think will be most interesting to you.

. . . . . .

'I am ashamed, dear B., to send you this imperfect account of so valuable a paper, but I trust that my willingness to give you news will plead my excuse. I hope you will soon read a copy of the paper at large, and have no doubt you will perceive on every page the inquiring spirit of the author.

. . . .

'I must not forget the proof you have given me of your feelings, truly of friendship, in the dilemma, and I am extremely sorry that I should in any way have occasioned you embarrassment. I am indebted much to you for your care in concealing such things as you supposed I intended for you alone. They were written for you alone; but, at the same time, I did not wish that my mother should remain ignorant of them. I have no secrets from her, and it was the insignificance alone that made me quiet on the subject. I would rather my mother should see or hear the first sheet of this paper than otherwise, for where the causes are open, the conduct can be better judged of. With this part you can do as you please; but there is as yet little in it can interest her, and I do not know that I shall add much more

'I must, however, tell you that we are in the midst of the Carnival, a scene of great mirth and jollity amongst the Romans.

. . . . . .

'I went this morning to a masquerade ball, between two and five o'clock, and found it excellent.

'Now for news!!! We shall part in a few weeks (pray write quickly) for Naples, and from thence proceed immediately to Sicily. Afterwards our road is doubtful; but this much I know, that application is made for passports to travel in the Turkish Empire, and to reside in Constantinople; that it is Sir Humphry's intention to be amongst the Greek Islands in March, and at Athens early in the spring. Thus you see, B., a great extension is made to our voyage—an extension which, though it promises much novelty and pleasure, yet I fear will sadly interrupt our correspondence. Have the goodness, therefore, to write quickly, and tell all my friends that you can to do the same, or I shall not get the letters. I shall make a point of writing to you as long and as late as I can.

'I will not pretend to know whether it is time to leave off or not, but I think it is impossible for you to get through this letter of twelve pages in less than three or four readings. How it has got to such a length I know not, for I have as yet read no part of it over; and even now I find I could write you a long letter, were this and the subjects of it annihilated; but I must cut it short. Pray remember me with the strongest affection to my mother and friends, and to your family. (Excuse the repetition.)

. . . . .

'Adieu, dear friend. With you I have no ceremony;

the warmest wishes that friendship can dictate are formed for you by

'M. FARADAY.

. . . . .

'Le donne Italiane sono sfacciate, pigrissime e sporchissime. Come dunque volete fare una comparazione fra loro e l' Inglese? Addio, caro amico!'

Monday, 30th.—Went in a domino to the mask ball this morning, and was much amused, though there were but few people, and the greater number were in their common day-dress. The theatre in which it was held was a very fine one, large and in excellent condition, and extremely well lighted. A vast number of chandeliers were suspended from all parts of the roof and filled with wax-candles, and every box was also lighted up. The stage and the pit were thrown together by a flight of steps. The pit was given for waltzing and the stage for cotillon and country dances, and two good bands of music were employed in the theatre. Other rooms in the wings were thrown open, some for dancing and some for refreshments. The three lower tiers of boxes were shut, but the rest were open to the maskers and the people in the house. A guard of soldiers was placed in the house to preserve order, and a gentleman in black with a cocked hat sat in the centre box and overlooked the whole. He appeared to enjoy the scene very slightly, and was, I suppose, there as fulfilling a duty in looking over the whims of the place.

In the afternoon there was much masking in the Corso, and the sugar plums, which were only seen in the sellers' baskets on the first evening, were now flying in the air. These *confetti*, as they are called, are merely plaster or old mortar broken into small pieces, and dropped in a mixture of whiting; but the men take care to sell them dear, though the price generally depends upon the eagerness of the purchaser at any moment to have them. With these the battles are carried on between mask and mask, or between carriage and carriage. None but masks are allowed to throw, though this rule is transgressed from every window. The most dreadful contests are carried on between the carriages as they pass each other, and I found the English were much more eager at this sport than the Romans. I know an English window from which eight crowns' worth of *confetti* were thrown this afternoon.

In my way to the Academy Lanesi I made a great blunder — I mistook a burial for part of the masquerade!!! But from the habits of the priests and mourners who attended it, it might be thought the mistake was theirs who put religion in those things, rather than mine who took it for masking. Their sackcloth coats, very similar to what the masked clowns and punchinellos wear, their enormous knotted cords tied round their waists, their sandals, and their caps, like a brewer's straining-bag, with two little holes for the eyes, were as complete a mask as it is possible to make; and it was not till by chance I saw the body that I thought it was a serious affair.

Wednesday, February 1st.—Experiments at home all day on a new solid compound of iodine and oxygen, which Sir Humphry discovered on Monday. It is formed by the action of euchlorine on iodine, which produces at the same time the compound of iodine and chlorine, and of iodine and oxygen. Its properties are many and curious, and it has enabled Sir H. to demonstrate (abroad) the truth of his ideas respecting iodine and its various compounds and combinations.

Monday, 6*th*.—Went to this morning's masked ball in a domino, and found it very full; as no one knew me, at least for some hours, I amused myself a good deal with such as I was acquainted with. I stopped there till daylight, and then came home.

Tuesday, 7*th*.—To-day is the last of the Carnival, and all Rome swarmed with masks: they were in every corner, just as you find the fleas there, and the quantity of *confetti* thrown away was astonishing. A race of nine horses cleared the Corso for a short time; but as if really to give a long adieu to the season, the carriages and masks entered again. They were, however, restrained from paying the last ceremonies to their departing pleasures, for the Pope would not allow of what had always taken place till this year. It had been the custom of the masks to promenade, on this the last day, with lighted tapers in their hands, crying out *Mort' è Carnevale*; but now that the Pope himself was at Rome, he would not allow of such a mockery of their burial service, which they wish to have considered as something serious.

To-night's ball was the last of the profane pleasures the season allowed them, and indeed it was well enjoyed. I found all Rome there, and all the English besides. It was too full for dancing, and the amusement was principally the jokes of those that were not known to those whom they knew. I was in a nightgown and nightcap, and had a lady with me whom I had not seen till that night, but who knew all my acquaintances; and between us we puzzled them mightily, and we both came away well entertained.

TO MR. R. ABBOTT.

'Rome: January 12, 1815. Received February 5.

'Dear Friend,—I hasten to make use of another opportunity, which the kindness of Sir Humphry Davy offers to me, to pay you a letter which I have long but unwillingly owed you.

. . . . . .

'Rome is far more amusing, pleasant, and interesting now than it was the last time we were here. We have now swarms of English about us, who keep this part of the world constantly in motion. The season is more interesting, the weather is very mild and fine, and the Carnival approaches; added to which, time has added a little more to my stock of Italian, and I find myself more capable of searching out and inquiring for things and information.

. . . . . .

'It happened, about three weeks ago, that a senator was elected; and upon the addition of a member to that august body, the Senate of Rome, it was said that fine doings would take place. A procession was promised to please the mob, and give them a high opinion of their new director; but as the weather happened to be bad on the day fixed, it was very unceremoniously put off as a thing of little importance. Now this opinion of it may be very correct, for I should think 'twas of no importance at all, but for such an arrangement to be altered, which had been made by the government, tended to give very light ideas of the government itself. The procession, however, took place on the first day of the year, when the weather was beautiful, and the town shone forth in great splendour. In the morning, preparations were made by spreading mould along those streets through which the procession was likely to pass in its way from Monte Cavallo, the Pope's palace, to the Campidoglio, or Capitol, where the senate house stands. About twelve o'clock, the fronts of the houses in those streets were highly decorated by tapestry and hangings suspended from the windows, many pieces of which had moved from the floors, and many from the beds. About three o'clock, the procession moved, and made a pretty sight enough, but certainly not what I expected for a Roman senator: it was clean and in good order, but short, and neither the Pope nor the cardinals were there.

. . . . . .

'Yours ever and sincerely,

'M. FARADAY.'

Saturday, 11*th*.—Experiments at home on the new compound of oxygen and chlorine which Sir Humphry discovered a few days ago. It is a gas of a very bright greenish colour, which detonates into chlorine and oxygen by a heat a little above that of boiling water. It was detonated in a comparative experiment against chlorine. One volume increased to nearly $1\frac{1}{2}$ in both experiments; but the products of the decomposed new gas contained 1 of oxygen and $\frac{1}{2}$ of chlorine, and the products of the detonated euchlorine were 1 of chlorine and $\frac{1}{2}$ of oxygen. A small piece of phosphorus introduced into it caused a spontaneous detonation. A solution of it, by its action on solution of the alkalies, gradually formed hyperoxymuriates, &c.

FARADAY TO HIS MOTHER.

'Rome: February 13, 1815.

'My dear Mother,— . . . . I wrote lately to B., and put the letter in the post: I do not know when he will get it, or whether he will get it at all; but if he does, he will tell you that we are going to Greece and Turkey immediately. I thought we were going there, but at present things seem a little more unsettled. We go shortly to Naples, and, if we can, from Naples to Sicily; afterwards I know not what road we shall take: perhaps we shall go immediately by water to the Archipelago or Grecian Islands, or perhaps we may return, up Italy again, across the Alps, see Germany, and then pass, by Carynthia, Illyria, and Dalmatia, into Turkey. Things being in this state, I can say nothing more particular about the road at present, though I can tell you to a moral certainty that we are to see Constantinople.

'The mention of these places calls England to my mind, now farther from me than any of them; and much as I wish to see these places, yet the idea of England fills my mind, and leaves no room for thoughts of other nations: 'tis still the name which closes the list, and 'twill ever be the place I am desirous of seeing last and longest. Our travels are amusing and instructive, and give great pleasure; but they would be dull and melancholy indeed if the hope of returning to England did not accompany us in them. But however, dear Mother, circumstances may divide us for a time, and however immense the distance may be between us, whatever our respective states may be, yet never shall I refrain from using my utmost exertions to remind you of me. At that distance to which we may

1815. ÆT. 23.

go, I shall despair of hearing from you; but if it is possible, my letters shall find you out, and I trust you will never be tired of them.

. . . .

'Give my kindest love to —— and ——, and my remembrances to all who ask you of me. And believe me, dear Mother, ever your most sincere and affectionate son,

'M. FARADAY.'

FARADAY TO HUXTABLE.

'Rome: February 13, 1815.

'Dear Huxtable,— . . . . . As for me, like a poor unmanned, unguided skiff, I pass over the world as the various and ever-changing winds may blow me; for a few weeks I am here, for a few months there, and sometimes I am I know not where, and at other times I know as little where I shall be. The change of place has, however, thrown me into many curious places and on many interesting things; and I have not failed to notice, as far as laid in my power, such things as struck me for their importance or singularity. You will suppose that Sir H. Davy has made his route as scientific as possible, and you must know that he has not been idle in experimental chemistry; and, still further, his example did great things in urging the Parisian chemists to exertion. Since Sir H. has left England, he has made a great addition to chemistry in his researches on the nature of iodine. He first showed that it was a simple body. He combined it with chlorine and hydrogen, and latterly with oxygen, and thus has added three acids of a new species to the science. He combined it with the metals, and found a class of salts analogous to the hyperoxymuriates. He still further combined these substances, and investigated their curious and singular properties.

'The combination of iodine with oxygen is a late discovery, and the paper has not yet perhaps reached the Royal Society. This substance has many singular properties. It combines both with acids and alkalies, forming with acids crystalline acid bodies; and with the alkaline metal oxyiodes, analogous to the hyper-oxymuriates. It is decomposed, by a heat about that of boiling oil, into oxygen and iodine, and leaves no residuum. It confirms all Sir H.'s former opinions and statements, and shows the inaccuracy of the labours of the French chemists on the same subjects.

'Sir Humphry also sent a long paper lately to the Royal Society on the ancient Greek and Roman colours, which will be worth your reading when it is printed; but if you please, for present satisfaction, Mr. B. Abbott can and will, I have no doubt, with pleasure read you a short account of it.

'Sir H. is now working on the old subject of chlorine, and, as is the practice with him, goes on discovering. Here, however, I am not at liberty to say much, but you may know that he has combined chlorine and oxygen in proportions differing from those of euchlorine. The new substance is a very beautiful yellow-green-coloured gas, much deeper than euchlorine. It explodes when heated with a sharp report, and 1 volume gives 1 of oxygen and ½ chlorine nearly; whereas 1 of euchlorine gives 1 of chlorine and ½ of oxygen: so that the new gas contains four times as much oxygen to the same volume of chlorine that euchlorine does.

'I beg to be excused for thus intruding subjects

1815. ÆT. 23.

which, perhaps, now have no charm for you, for your time I suppose is filled with medicine; but I hope you will attribute it to my wish to give a little value at least to my letter; and in whatever way you may receive it, I will still maintain that Sir H.'s discoveries are valuable. But I find my time runs short, though my subjects are not yet exhausted.

. . . .

'I present, with the certainty of their being accepted, the best wishes of yours, ever sincerely,

'M. FARADAY.'

FARADAY TO ABBOTT.

'Rome: February 23, 1815.

'Dear B——, In a letter of above twelve pages I gave answers to your question respecting my situation. It was a subject not worth talking about, but I consider your inquiries as so many proofs of your kindness and the interest you take in my welfare, and I thought the most agreeable thanks I could make you would be to answer them. The same letter also contained a short account of a paper written by Sir Humphry Davy on ancient colours, and some other miscellaneous matters.

'I am quite ashamed of dwelling so often on my own affairs, but as I know you wish it, I shall briefly inform you of my situation. I do not mean to employ much of this sheet of paper on the subject, but refer you to the before-mentioned long letter for clear information. It happened, a few days before we left England, that Sir H.'s valet declined going with him, and in the short space of time allowed by circumstances another could not be got. Sir H. told me he was very sorry, but that, if I would do such things as were absolutely necessary for him until he got to Paris, he should there get another. I murmured, but agreed. At Paris he could not get one. No Englishmen were there, and no Frenchman fit for the place could talk English to me. At Lyons he could not get one; at Montpellier he could not get one; nor at Genoa, nor at Florence, nor at Rome, nor in all Italy; and I believe at last he did not wish to get one: and we are just the same now as we were when he left England. This of course throws things into my duty which it was not my agreement, and is not my wish, to perform, but which are, if I remain with Sir H., unavoidable. These, it is true, are very few; for having been accustomed in early years to do for himself, he continues to do so at present, and he leaves very little for a valet to perform; and as he knows that it is not pleasing to me, and that I do not consider myself as obliged to do them, he is always as careful as possible to keep those things from me which he knows would be disagreeable. But Lady Davy is of another humour. She likes to show her authority, and at first I found her extremely earnest in mortifying me. This occasioned quarrels between us, at each of which I gained ground, and she lost it; for the frequency made me care nothing about them, and weakened her authority, and after each she behaved in a milder manner. Sir H. has also taken care to get servants of the country, ycleped *lacquais de placé*, to do everything she can want, and now I am somewhat comfortable; indeed, at this moment I am perfectly at liberty, for Sir H. has gone to Naples to search for a house or lodging to which we may follow him, and I have nothing to do but see Rome, write my journal, and learn Italian.

'But I will leave such an unprofitable subject, and tell you what I know of our intended route. For the last few weeks it has been very undecided, and at this moment there is no knowing which way we shall turn. Sir H. intended to see Greece and Turkey this summer, and arrangements were half made for the voyage; but he has just learned that a quarantine must be performed on the road there, and to this he has an utter aversion, and that alone will perhaps break up the journey.

. . . . . .

'Since the long letter I wrote you, Sir H. has written two short papers for the Royal Society—the first on a new solid compound of iodine and oxygen, and the second a new gaseous compound of chlorine and oxygen, which contains four times as much oxygen as euchlorine.

'The discovery of these bodies contradicts many parts of Gay-Lussac's paper on iodine, which has been very much vaunted in these parts. The French chemists were not aware of the importance of the subject until it was shown to them, and now they are in haste to reap all the honours attached to it; but their haste opposes their aim. They reason theoretically, without demonstrating experimentally, and errors are the result.

'I intended at first to give you some account of waterfalls in this sheet, but I fancy only the name will be seen, for (not I, but) the bearer of this letter has no more time to allow me.

. . . . . .

'I am, my dear Friend, yours ever and faithfully,

'M. FARADAY.'

Friday, March 3rd.—Left Tondi; the first two stages rode a saddle-horse. Now, though I am no rider, yet the circumstance must not be attributed to me alone that the horse and I were twice heels over head, but rather it is a wonder that it did not happen oftener in nine miles. A tailor would have said that the horse was religious, and that it only did as other Italians do when they grow old and feeble; but that did not satisfy me, and I would rather have had a beast that would have gone on orderly upon his legs.

Tuesday, March 7th.—I heard for news that Bonaparte was again at liberty. Being no politician, I did not trouble myself much about it, though I suppose it will have a strong influence on the affairs of Europe.

Thursday, 16th.—I intended this morning to dedicate this day to Pompeii, but on Sir Humphry's asking me whether I would go with him to Monte Somma, I changed my mind. We left Naples about 10.30, and took the usual road to Resina. The weather was clear, the atmosphere heavy, the wind fair for ascending Vesuvius, which rose before us as the gates of another world, and was still marked by yesterday's fall of snow. At Resina we bought bread and oranges, and then began the ascent. After an hour spent amongst the vineyards we came upon the plain of lava, and crossed on towards Monte Somma. Here the guide pointed out some particular stream of this matter. We crossed the lava of 1814. Its surface was very rough and craggy, as if it had issued out in a state of imperfect fusion and almost ready to become solid at the moment, yet it was not more than eleven or twelve inches thick. The old lava (17..?) had a very different appearance. It had been, if not more fusible, more fluid, and had

1815. ÆT. 23.

flowed flatter and smoother. It was even thinner than the former.

We now began to ascend the hill of the Hermitage, and here the guide pointed out cinders or ashes of the same kind, and of the same shower, that overwhelmed Pompeii; and it is of these, and of volcanic products more ancient, which Sir Humphry gave to a period of time long before Pliny, that this hill is formed. We made no stop at the Hermitage, except to view the plain which we had just left below us, and to note the directions of the currents. Here the guide again reverted to the old lava of 17..? and related in what manner it was approaching to Naples itself, and the rapid advances it made every hour, insomuch that it had soon passed many other streams of lava, and begun to menace the city, when good St. Januarius went to it, and standing before it with a crucifix in his hand, he raised it in front of the burning river, and bade it stop, which it immediately did, and became firmly fixed in the same place. It is not always, however, that the image or saint possesses such virtue or faith, and then curious contests arise between the people and their ineffectual protector, and even the Virgin Mary has been so much abused as to have phlegm thrown in her face.

After a little further progress I left Sir Humphry, for he intended to see Monte Somma only, and I wished to ascend the cone of Vesuvius. I therefore continued my way alone along the lake of lava, and soon gained the foot of the cone, where I found several asses and eight men who belonged to a company then up the mountain. On hearing that the state of things was altered above, and was no longer as when I saw it before, I took a guide with me, and began to climb. After a little while

I saw the company above me just coming over the edge of the hill, and it was at this moment I gained a correct idea of the size of this ash hill. From the uniformity of its figure and inclination, the eye is deceived, and thinks it much smaller than it really is ; but when I saw some moving spots at the summit, and, by the guide's assistance, distinguished them to be men, I was aware in some measure of the immense space between them and me. We continued to ascend regularly, except at intervals when we turned round to enjoy the fine view from this elevated spot. The company now approached us, sending down a shower of stones before them. The ascending and descending path is different, so that no one is in danger from a blow, except such as are first of the descending corps, but in them inattention would be dangerous. In thirty minutes we gained the edge of the crater, and got on flatter, i.e. less inclined ground, for as to smoothness, it was much rougher than the hill. The fatigue of the ascent in a hot sunshine had made me very thirsty. With pleasure I ate with my hand some of the still unmelted snow on the mountain.

This uneven surface presents many spots where the vapours and smoke issue out, sometimes even from the centre of rocks of lava ; they rise dense and heavy, and appear to be sulphurous and carbonic acids, with water, in a state of vapour. Their effects upon the lava were to bleach it, making it of a fine white or yellow colour, and in many spots they deposit muriate of iron and muriate of ammonia. The ground is very wet in many places from the melting of the snow and the condensation of the vapours that rose from the interior, and the guide, from that circumstance, promised me a fine view of the crater. After about ten minutes' further progress we came to an elevated mass of lava, and from thence saw the crater about sixty yards in advance; but here we stopped awhile to see it at this distance. The scene was grand in the extreme, and cannot be conceived but from the seeing of it. The cloud of smoke rose very rapid and high in the atmosphere, and moved off in a side direction, so as to leave us without fear of being annoyed by it. The colouring of the place was very strange, though brilliant. The smoke at moments took various tinges from the sun on the part opposed to its rays, and the opposite side of columns possessed all the sombre black and waving red hues of that which might be supposed to issue from the abyss. The dark burnt ground was irregularly arrayed in many colours of the greatest beauty, but they struck the eye as being unnatural. The yellows were muriate of iron and lava, with various tints from its natural black colour to white, according to the time or the power with which the sulphurous vapours had acted on it. The reds and greens were mixtures of the bleached lava with iron. No sulphur was present in a concrete form, and no smell of sulphur except from the vapours of the volcano. The general smell of the place was like chlorine.

From the spot that we now occupied I heard the roar of the fire, and at moments felt the agitation and shakings of the mountain; but the guide, not satisfied with this, went forward, and we descended some rocks of lava and proceeded onwards towards the very edge of the crater, leaping from one point to another, being careful not to slip, not only to avoid the general inconveniences of a fall, but the being burnt also, for at the bottom of a cavity the heat was in general very great. I had nearly, however, been down, for, whilst stepping, skipping, &c., the guide suddenly cried out to look, and I did so, though falling. I saw a large shower of red-hot stones in the air, and felt the strong workings of the mountain; but my care was now to get to the crater, and that was soon done. Here the scene surpassed everything. Before me was the crater, like a deep gulf, appearing bottomless from the smoke that rose from below. On the right hand this smoke ascended in enormous wreaths, rolling above us into all forms; on the left hand the crater was clear, except where the fire burst out from the side with violence, its product rising and increasing the volume of volatile matter already raised in the air. The ground was in continual motion, and the explosions were continual, but at times more powerful shocks and noises occurred; then might be seen rising high in the air numbers of red-hot stones and pieces of lava, which at times came so near as to threaten us with a blow. The appearance of the lava was at once sufficient to satisfy one of its pasty form. It rose in the air in lumps of various size, from $\frac{1}{2}$ lb. to 25 lbs. or more. The form was irregular, but generally long, like splashes of thick mud; a piece would oftentimes split into two or more pieces in the air. They were red-hot, and, when they fell down, continued glowing for five, ten, or fifteen minutes. They generally fell within the crater, though sometimes a piece would go far beyond its edge. It appeared as if splashed up by the agitation of a lake of lava beneath; but the smoke hid all below from sight. The smoke generally rose in a regular manner, and, though the noises, explosions, and trembling varied much, yet the cloud seemed to rise with the same strength and impetuosity. I was there, however, during one explosion of very great force, when the ground shook as with a strong earthquake, and the shower of lava and of stones ascended to a very great height, and at this moment the smoke increased much in quantity. The guide now said this place was not safe, from its exposed situation to the melted lava and to the smoke, and because it oftentimes happens that a portion of the edge of the crater is shaken down into the gulf below. We therefore retreated a little, and then sat down, listened, and looked.

After a while we returned. This was rapid work, but required care, from the heat of the lava and the chance of a fall of some yards. The descent of the cone is made over the softest part, that which is most equally spread with ashes and in the finest form. Every step is worth twenty of the ascending ones, and it took us four minutes and a half to return over a space which occupied in ascending thirty-five minutes of our time. At the Hermitage I found some acquaintance, but not Sir Humphry, and I therefore continued to descend, and got to Resina in good time.

I was very glad of this opportunity of ascending Vesuvius; for I had heard so many and such different accounts from persons who have lately been up it, that I thought it must be in a very changeable state, or, at least, that it had changed much since I saw it last. This it certainly has done, and by to-day's walk I have gained a much clearer idea of a volcano than I before possessed.

Monday, March 21st.—We left Naples this morning at five o'clock. The weather pleasant, but cool. We pressed forward the whole of the day, and fully employed it in getting to Terracina. In passing Fandi we were saluted by the Neapolitan troops, who were coming into the town from their recreation by the gate we wanted to go out at. Their salutations were stares, hurras, hisses, groans, laughing and chattering, and all apparently for want of something to do. There were a great number of them here, and the town was more than full with them, but for what purpose they are here is as yet unknown.

Wednesday, 23*rd*.—Rome. The Romans are now in much agitation respecting the motions of Murat, and made eager inquiries about his advance, &c., but they made no preparations to oppose him. The Pope goes to-day from this place, and the Cardinals will all be off in two or three days more. No post-horses to be had.

Thursday, 24*th*.—We wanted to go to-day, but finding that everybody else wanted to go too, and that no horses were to be had, we were obliged to delay a little our departure. Everything, however, was packed up, and every means used to set us and our luggage in motion. At last carriage-horses were hired at an immense expense to take us to Cività Castellana to-morrow.

Wednesday, *March* 30*th*.—Mantua is singularly twisted by fortifications and earthworks, and labourers are now employed by hundreds to twist it still more by the same kind of arrangements. Indeed, everything is prepared, and everybody is preparing, for war. Mantua has on this side a very pretty and picturesque appearance, but this I fancy it owes in a great measure to the magnificent background given to it by the Alps. We wished to get through the town as soon as possible, but were destined to remain in it some hours. We found some trouble in getting into it, and we found still more in getting out of it. The passport was asked for at the outer gate; it was taken into the bureau, examined, and registered. It was then, at the distance of twelve or fourteen yards, asked for at the inner gate, examined and registered as before, and then sent by a soldier to the police office. In the meantime we were permitted to proceed to the post-house, and there remained. After a while the soldier came back, and said Sir Humphry must go to the police-office to answer some questions. In about forty minutes Sir Humphry came back with the permission, but it was not good. It had been signed by the police only, and not by the commandant. It was to go back again, and I went with it. At the police-office I found them examining the passport, and I witnessed the several tedious operations of examination, registering, signing, and sealing. The passport then passed the ordeal of their hands, and then we got it re-signed or countersigned by the commandant. All was now valid, and we got the horses and considered ourselves in a state of motion, but we found it to be intermitting only, and not continual. At the gate of exit we were stopped, the passport examined, registered, &c., and the same done at another bureau about twelve yards in advance. During these examinations the traces of a wagon laden with hay broke just as it got on the drawbridge, and we had to wait until they pleased to mend them. At last we saw the outside of the town, having, much against our will, remained two hours and a half in it.

. . . . . .

FARADAY TO HIS MOTHER.

'Bruxelles: April 16, 1815.

'My very dear Mother,—It is with no small pleasure I write you my last letter from a foreign country, and I hope it will be with as much pleasure you will hear I am within three days of England. Nay more, before you read this letter I hope to tread on British ground, but I will not make too sure, lest I should be disappointed; and the sudden change and apparently termination of our travels is sufficient to remind me that it may change again. But, however, that is not at all probable, and I trust will not happen.

'I am not acquainted with the reason of our sudden return; it is, however, sufficient for me that it has taken place. We left Naples very hastily, perhaps because of the motion of the Neapolitan troops, and perhaps for private reasons. We came rapidly to Rome, we as rapidly left it. We ran up Italy, we crossed the Tyrol, we stepped over Germany, we entered Holland, and we are now at Brussels, and talk of leaving it to-morrow for Ostend; at Ostend we embark, and at Deal we land on a spot of earth which I will never leave again. You may be sure we shall not creep from Deal to London, and I am sure I shall not creep to 18 Weymouth Street; and then—— but it is of no use. I have a thousand times endeavoured to fancy a meeting with you and my relations and friends, and I am sure I have as often failed: the reality must be a pleasure not to be imagined nor to be described. It is uncertain what day we shall get to London, and it is also uncertain where we shall put up at. I shall be thankful if you will make no inquiries after me anywhere, and especially in Portland Place, or of Mr. Brande. I do not wish to give occasion for any kind of comments whatever on me and mine. You may be sure that my first moments will be in your company. If you have opportunities, tell some of my dearest friends, but do not tell everybody—that is, do not trouble yourself to do it. I am of no consequence except to a few, and there are but a few that are of consequence to me, and there are some whom I should like to be the first to tell myself—Mr. Riebau for one. However, let A. know if you can.

'I come home almost like the prodigal, for I shall want everything.

. . . . . .

'I cannot find in my heart to say much here to B. and R., because I want to say it myself, and I feel that I am too glad to write it. My thoughts wander from one to another, my pen runs on by fits and starts, and I should put all in confusion. I do not know what to say, and yet cannot put an end to my letter. I would fain be talking to you, but I must cease.

'Adieu till I see you, dearest Mother; and believe me ever your affectionate and dutiful son,

'M. FARADAY.

''Tis the shortest and (to me) the sweetest letter I ever wrote you.'

The feeling that bursts out in his letter to his friend Abbott from Rome, and in this last letter to his mother from Brussels, contrasts most remarkably with the tone of his journal. Both are strikingly characteristic of Faraday—the journal, by the absence of every word of gossip, and by the keenness of his remarks on everything that came before him—the letters, by their kindness, which seems often too much to find utterance in words.

CHAPTER IV.

EARLIER SCIENTIFIC EDUCATION AT THE ROYAL INSTITUTION—FIRST LECTURES AT THE CITY PHILOSOPHICAL SOCIETY—FIRST PAPER IN THE 'QUARTERLY JOURNAL OF SCIENCE.'

1815. ÆT. 23.

ON May 7, a fortnight after his return to England, Faraday was engaged at the Royal Institution as assistant in the laboratory and mineralogical collection, and superintendent of the apparatus, with a salary of thirty shillings a week. Apartments were also granted to him; but a month passed before he was put in possession of them. Up to this time the love which Faraday had for knowledge, and his earnest search for it, are to be seen in every letter he wrote and in everything that he did. This is the moving force which led to most of his actions, and occasionally it bursts out in words like these—'Trade which I hated, and science which I loved;' 'I almost wished that I had been insulated and alone, that I might have accepted Sir H. Davy's offer without a regret at leaving home.' In another letter he wrote, 'The glorious opportunity I enjoy of improving in the knowledge of chemistry and the sciences with Sir H. Davy:' and with that innate humility which was increased by his religion, he said, 'I have learned just enough to perceive my ignorance;' 'The little knowledge I have gained makes me wish to know more.'

1815. ÆT. 23.

This was the state of his mind when he returned to the place where he was to be further educated for his great scientific work. Faraday had now full knowledge of his master's genius and power. He had compared him with the French philosophers whilst helping him in his experiments on iodine; and he was just about to see him engage in those researches on fire-damp and flame, which ended in the glorious invention of the Davy lamp, a discovery which gave to Davy a popular reputation, even beyond that which he had gained in science by the greatest of all his discoveries—potassium.

The care with which Faraday has preserved every note-book and manuscript of Davy's at the Royal Institution, the remarks regarding Davy in his letters, the earnestness of his praise of Davy's scientific work, show that he fully acknowledged all the debt which he owed to his master. But, with all his genius, Davy was hurt by his own great success. He had very little self-control, and but little method and order. He gave Faraday every opportunity of studying the example which was set before him during the journey abroad, and during their constant intercourse in the laboratory of the Royal Institution; and Faraday has been known to say that the greatest of all his great advantages was that he had a model to teach him what he should avoid.

The rapid progress that he made in his self-education during the first five years of his fixed abode at the Royal Institution is well seen, first, in the lectures which he gave; secondly, in the entries which he made in his commonplace-book; thirdly, in the papers which he published; and, lastly, in the letters which he wrote.

Under these different heads, as far as possible in his own words, year by year the history of his life will be told.

In his works he will show his own growth in science, and in his letters he will set forth his own character.

I.

It 1816, the state and progress of his knowledge is seen chiefly in the lectures which he gave at the City Philosophical Society. His first lecture was on January 17. His subject was the general properties of matter. In the course of the year he gave six more lectures: these were—1, On the Attraction of Cohesion; 2, On Chemical Affinity; 3, On Radiant Matter; 4, 5, and 6, On Oxygen, Chlorine, Iodine, Fluorine, Hydrogen, and Nitrogen.

These were Faraday's earliest lectures. He wrote them out with great care; and it is interesting now to see in his own words the views which he then held regarding the unity, the relationships, and the nature of matter and force—subjects which continued to occupy his thoughts almost to the end of his life.

In his first lecture, on the General Properties of Matter, he says:—

'With much diffidence I present myself before you this evening as a lecturer on the difficult and refined science of chemistry—a science which requires a mind more than mediocre to follow its progress; but I trust that my efforts to fulfil my duty as a member of this Society will be received favourably, though I may fail in them.

'Chemistry is a knowledge of the powers and properties of matter, and of the effects produced by those powers; but it is evident that we can only become acquainted with these as we become acquainted with matter; and, *vice versâ*, we can only know matter as we know its properties. It would seem right, therefore, that we should speak of them in connection; but our knowledge becomes much more clear, precise, and orderly, if we divide it here into two parts—the one of the properties of matter, and the other of matter itself, or rather of its varieties.

'But, as I have said before, we can only know the properties of matter by investigating matter itself; and we have no other means of distinguishing matter than by its different properties. It seems, therefore, difficult to separate them even in idea; and if we separate them, the question arises, which should be first described? But a little attention will point out to us that division and arrangement which appear most natural, and which can be most easily retained in the mind.

'We are all able in some degree to form ideas of the properties of matter abstracted from matter itself, and can discuss the phenomena of attraction or repulsion without including the idea of any particular substance; but matter cannot be described except by its properties, nor distinguished but as these differ from the properties of other matter either in kind or degree. It would be impossible for me to describe sulphur or charcoal if I refused to mention their properties, or to distinguish them from each other unless I said in what properties they differed. I shall, therefore, first endeavour to enumerate and illustrate the properties of matter.'

. . . . . .

In his third lecture, on Chemical Affinity, he said:—

'Thus far our attention has been given to the phenomena produced by the efficient exertion of chemical affinity. We have gained a knowledge of the varieties of action and effect produced by degrees in its power and the interference of cohesion. That this power is an adherent property in bodies, and causes them to approach each other, is evident from every phenomenon connected with it that has been made known. But this knowledge is not sufficient to satisfy the curiosity of man; and in his restless desire after further information he pants to become acquainted with the cause of this power. This thirst for knowledge has induced many to torture nature in hopes to discover her secrets; and though their labours have not been repaid with anything like the success desired (at least on this point), yet they have given much extraordinary and interesting information; and it shall now be my object to detail to you such part of this new knowledge as will throw light on the nature of chemical affinity.

'In latter times, which are most important as relating to this power, the researches of philosophers have been directed to the investigation of the influence exerted by electricity over chemical affinities, as being the most promising source of information. I have in a former lecture stated to you the discovery of that influence, and described some of its effects; but it is now my duty to point out more clearly the close connection which exists between the electric power and that of chemical affinity, and to lay before you those propositions relative to the subject which present themselves in the labours of modern philosophers.'

. . . . . .

In his fourth lecture, upon Radiant Matter:—

'Assuming heat and similar subjects to be matter, we shall then have a very marked division of all the varieties of substance into two classes: one of these will contain ponderable and the other imponderable matter.

'The great source of imponderable matter, and that which supplies all the varieties, is the sun, whose office it appears to be to shed these subtle principles over our system.

. . . . . .

'The metals are among the most opaque bodies we are acquainted with, yet when beat into very thin leaves they suffer light to pass. Gold, one of the heaviest of the metals, when beaten out and laid upon glass, forms a screen of much transparency, and anything strongly illuminated, as by the sun, may be seen through it. It has been said that this is occasioned by the existence of small holes in the leaves, which permit the light to pass through them, and that it does not pass through the body of the metal. If by small holes be meant the pores of the metal, the explanation will readily be granted: but then the metal must be considered to a certain degree transparent, for the transmission of light through pores is the only way in which it can be transmitted at all, and nothing else takes place in transparent bodies; but if it be said that the existence of such holes as the light is supposed to pass through is accidental, and only happens when the leaves are made very thin, then arguments can be opposed to such a statement; for supposing it to be true, the light which passes should be white, whereas it is coloured, and the colour is found to depend on the metal being influenced by other substances which it may contain. Pure gold appears by transmitted light of a purple colour; gold with a little silver, bluish; with a little copper, green; with iron, red; and these changes of colour almost prove that light does not pass through such small accidental holes, but actually through the pores of the metal, as with other transparent matter.[1]

. . . . . .

'The conclusion that is now generally received appears to be that light consists of minute atoms of matter of an octahedral form, possessing polarity, and varying in size or in velocity.

. . . . . .

'If now we conceive a change as far beyond vaporisation as that is above fluidity, and then take into account also the proportional increased extent of alteration as the changes rise, we shall perhaps, if we can form any conception at all, not fall far short of radiant matter; and as in the last conversion many qualities were lost, so here also many more would disappear.

'It was the opinion of Newton, and of many other distinguished philosophers, that this conversion was possible, and continually going on in the processes of nature, and they found that the idea would bear without injury the application of mathematical reasoning—as regards heat, for instance. If assumed, we must also assume the simplicity of matter; for it would follow that all the variety of substances with which we are acquainted could be converted into one of three kinds of radiant matter, which again may differ from each other only in the size of their particles or their form. The properties of known bodies would then be supposed to arise from the varied arrangements of their ultimate atoms, and belong to substances only as long as their compound nature existed; and thus variety of matter

[1] Forty years after this lecture was given, Faraday published his last paper, full of experiments, in the *Philosophical Transactions*, upon this subject.

and variety of properties would be found co-essential. The simplicity of such a system is singularly beautiful, the idea grand, and worthy of Newton's approbation. It was what the ancients believed, and it may be what a future race will realise.'

In his fifth lecture, on Oxygen, Chlorine, Iodine, Fluorine, he said :—

. . . . . .

'Before leaving this substance, chlorine, I will point out its history, as an answer to those who are in the habit of saying to every new fact, "What is its use?" Dr. Franklin says to such, "What is the use of an infant?" The answer of the experimentalist would be, "Endeavour to make it useful." When Scheele discovered this substance it appeared to have no use, it was in its infantine and useless state; but having grown up to maturity, witness its powers, and see what endeavours to make it useful have done.

. . . . . .

'The third body we have to consider is called iodine, or iode. It was discovered in the year 1812, by M. Curtois, a saltpetre manufacturer at Paris.

'No suspicions were entertained at first of the true nature of this body. It gained but little attention, and was supposed to be a compound. A short paper was read on it to the Institute at Paris; but until Sir H. Davy hinted his suspicions that it was an undecomposed substance, it seemed to be gradually falling into neglect. During the winter of 1813, several papers on it by Sir H. Davy, Gay-Lussac, and others, appeared; and though in my own mind convinced to whom the merit of having first ascertained the true nature of this substance, its remarkable properties, and its compounds, belongs, it is not fit that I should influence you here. The public must judge from the different papers and their dates, and will if they discriminate correctly, give the merit where it is due.

. . . . . .

'It is probable that none of these bodies, oxygen, chlorine, iodine, and fluorine, are really simple in their nature.

'The cursory view which we have taken of them and of their history is sufficient to show the disadvantages attendant on ignorance and on undue faith in hypothesis.

'It may be observed how great are the alterations made in the opinions of men by the extension of inquiry; and this points out the imperfections of those originally held. Again, by adherence to a favourite theory, many errors have at times been introduced into general science which have required much labour for their removal. These circumstances are the cause of many obstructions in the path of knowledge. Whilst, however, we can thus observe those causes which have at former periods acted in a manner not agreeable to our wishes, let us be careful not to give occasion to future generations to return the charge on us. Man always forms opinions, and he always believes them correct. In pointing out the errors of another he endeavours to substitute for them his own views. At the present day we have our theories and laws, and we believe them to be correct, though they may probably fall, as others have done before them. 'Tis true that, warned by the example of others, we profess to be more reserved in our opinions, and more guarded in our decisions; and yet continual experience shows that our care applies rather to former errors than to those now likely to arise. We avoid those faults which we perceive, but we still fall into others. To guard against these requires a large proportion of mental humility, submission, and independence.

'The philosopher should be a man willing to listen to every suggestion, but determined to judge for himself. He should not be biassed by appearances; have no favourite hypothesis; be of no school; and in doctrine have no master. He should not be a respecter of persons, but of things. Truth should be his primary object. If to these qualities be added industry, he may indeed hope to walk within the veil of the temple of nature.'

In his sixth lecture, on Hydrogen, he said:—

'Although we should be able, from a knowledge of the importance of water, to form a very exalted idea of the value of hydrogen, as one of the elements which constitute it, yet it is by no means the highest point of view in which it can be placed before you. The attempt, indeed, to estimate the value of any one element in nature would be vain and presumptuous, for it is not possible that we can understand every use to which it may be and is constantly applied; yet still I think it is proper that we should evince our consciousness, as far as it extends, of the benefits we continually enjoy. Allow me, therefore, to point out to you the value of hydrogen, not only in water, but in all the common inflammable substances. Wood, coal, resin, wax, oil, and almost the whole of that variety of bodies which we use for the production of heat and light, contain hydrogen as an essential element; and these are valuable not only as sources of warmth and light, but for innumerable other uses which they possess in the hands of man. Look at them also in the hands of nature, see them formed from or forming the vegetable world, and then see them converted into other states and other compounds, and at last attaining to the important end of their creation in the composition of that masterpiece—man!'

II.

The entries which he made in his commonplace-book this year show what he did and what he read, what he thought, and even how he appeared at this time to one who at the City Philosophical Society had frequent and full occasions to know him well.

Among many notes, the most remarkable are on the production of oxygen; on mnemonics; on the combustion of zinc and of iron in condensed air; on mercury and tin; on crystals in oil of cassia; extracts from 'Rambler' and 'Spectator;' notes of a course of lectures on geology delivered at the Royal Institution by W. T. Brande, Esq., F.R.S.; tests for arsenic; analysis of native lime from Italy; comparison of French and English measures; an account of a visit to a silk-ribbon dresser; an account of Zerah Colburn, thirteen years old, the American calculating boy;[1] gauge for condensing apparatus; eudiometry; a long historical sketch; experiments on the absorption of nitrous acid by oil; two pieces of poetry—one on Love, and as this had some

[1] Sir H. Davy wrote to Faraday, 'Mr. Colburn, the father of the American boy who has such extraordinary powers of calculation, will explain to you the method his son uses in confidence: I wish to ascertain if it can be practically used.'

1816.
ÆT.24–25.

influence on his future life, it must be preserved; the other, called 'Quarterly Night,' October 2, 1816, was written by Mr. Dryden, a member of the City Philosophical Society.

What is the pest and plague of human life?
And what the curse that often brings a wife?
'Tis Love.

What is the power that ruins man's firmest mind?
What that deceives its host, alas! too kind?
What is't that comes in false deceitful guise,
Making dull fools of those that 'fore were wise?
'Tis Love.

What is't that oft to an enemy turns a friend?
What is't that promising never attains its end?
What that the wisest head can never scan,
Which seems to have come on earth to humble man?
'Tis Love.

What is't directs the madman's hot intent,
For which a dunce is fully competent?
What's that the wise man always strives to shun,
Though still it ever o'er the world has run?
'Tis Love.

Then show me love: howe'er you find it, 'tis still a curse,—
A thing which throws good sense behind it; sometimes much worse.
'Tis always roving, rambling, seeking t'unsettle minds,
And makes them careless, idle, weeping, changeful as winds.
Then come to me, we'll curse the boy the Cyprian goddess brought on earth;
He's but an idle senseless toy, and has no claim on manly worth.
The noble heart will ne'er resign Reason, the light of mental day,
Or idly let its force decline before the passions' boisterous sway.
We've honour, friendship, all the powers that still with virtue do reside;
They've sweetly strewed our lives with flowers, nor do we wish for aught beside.
Love, then, thou'st nothing here to do: depart, depart to yonder crew.

There is another entry on this subject in his note-book.

What is Love?—A nuisance to everybody but the parties concerned. A private affair which every one but those concerned wishes to make public.

A description of Faraday at this time is thus given in the 'Quarterly Night:'—

But hark! A voice arises near the chair!
Its liquid sounds glide smoothly through the air;
The listening Muse with rapture bends to view
The place of speaking, and the speaker too.
Neat was the youth in dress, in person plain;
His eye read thus, *Philosopher in grain*;
Of understanding clear, reflection deep;
Expert to apprehend, and strong to keep.
His watchful mind no subject can elude,
Nor specious arts of sophists ere delude;
His powers, unshackled, range from pole to pole;
His mind from error free, from guilt his soul.
Warmth in his heart, good humour in his face,
A friend to mirth, but foe to vile grimace;
A temper candid, manners unassuming,
Always correct, yet always unpresuming.
Such was the youth, the chief of all the band;
His name well known, Sir Humphry's right hand.
With manly ease towards the chair he bends,
With Watts's logic at his finger-ends.
'I rise (but shall not on the theme enlarge)
To show my approbation of this charge:
If proved it be, the censure should be passed,
Or this offence be neither worst nor last.
A precedent will stand from year to year,
And '*tis the usual practice* we shall hear.
Extreme severity 'tis right to shun,
For who could stand were justice only done?
And yet experience does most clearly show
Extreme indulgence oft engenders woe.
In striving then to hit the golden mean—
To knowledge, prudence, wisdom, virtue seen—

Let Isaac then be censured, not in spite,
But merely to evince our love of right.
Truth, order, justice, cannot be preserved,
Unless the laws which rule us are observed.
I for the *principle* alone contend,
Would lash the crime, but make the man my friend.'

. . . .

III.

His first original work was published this year, in the 'Quarterly Journal of Science.' It was an analysis of native caustic lime. In the volume of his 'Experimental Researches on Chemistry and Physics,' he has added a note:—'I reprint this paper at full length; it was the beginning of my communications to the public, and in its results very important to me. Sir Humphry Davy gave me the analysis to make as a first attempt in chemistry, at a time when my fear was greater than my confidence, and both far greater than my knowledge; at a time also when I had no thought of ever writing an original paper on science. The addition of his own comments, and the publication of the paper, encouraged me to go on making, from time to time, other slight communications, some of which appear in this volume. Their transferrence from the "Quarterly" into other journals increased my boldness, and now that forty years have elapsed, and I can look back on what successive communications have led to, I still hope, much as their character has changed, that I have not either now or forty years ago been too bold.'

Another mark of his progress appears in the fact that he mentions thus:—'When Mr. Brande left London in August, he gave the "Quarterly Journal" in charge to me: it has had very much of my time and care, and writing, through it, has been more abundant with me. It has, however, also been the means of giving me earlier information on some new objects of science.'

IV.

Lastly, from the letters which he wrote to his friend Benjamin Abbott, after he came from abroad, a glimpse of his life and character, during this year, can be obtained.

FARADAY TO ABBOTT.

'January 10, 1816.

'Dear A——, Many persons spend years in seeking honour, but still being unsuccessful call themselves miserable and unfortunate; but what are these cases to mine, who, when honour waits for admission, am obliged to refuse her entertainment? But so the fates, or the unlucky stars, or the gods, or something else, have decreed, and I am obliged to dissent from your arrangement for Thursday. It happens that my time for this week is completely cut up, and so that I cannot cut it over again. On Thursday evening I expect my old master, Mr. Riebau, at the Institution, and I shall be out both Friday and Saturday evenings.'

FARADAY TO ABBOTT.

'Friday evening (I believe), February 9, 1816.

'Dear A——, Be not offended that I turn to write you a letter because I feel a disinclination to do anything else, but rather accept it as a proof that conversation with you has more power with me than any other relaxation from business;—business I say, and believe it is the first time for many years that I have applied it to my own occupations. But at present they actually deserve the name, and you must not think me in laughing mood, but in earnest. It is now 9 o'clock, P.M., and I have just left the laboratory and the preparation for to-morrow's two lectures. Our double course makes me work enough, and to them add the attendance required by Sir Humphry in his researches; and then, if you compare my time with what is to be done in it, you will excuse the slow progress of our correspondence on my side. Understand me: I am not complaining; the more I have to do the more I learn; but I wish to avoid all suspicion on your side that I am lazy—suspicions, by the bye, which a moment's reflection convinces me can never exist.

. . . . . .

'Remember me to our friends, and believe me ever yours,

'M. FARADAY.'

FARADAY TO ABBOTT.

'September 23, 1816.

'Dear A——, . . . Whilst in the City I heard a curious charge from Mr. G., made by you against me, I suppose in joke; but given by him with so serious a face, that I was tempted to explain, a thing I rarely do to those who have no connection (necessarily) with my affairs. The charge was, that I deserted my old friends for new ones. Supposing that you intended this seriously, which I do not think you did, I shall take the opportunity to explain to *you* how my time is generally occupied. The duties of my situation (which is no sinecure) necessarily confine the time which I can dispose of to the evenings. Of these, Wednesday belongs to the Society; Saturday to Weymouth Street, generally; Monday and Thursday come into a system of instruction, and may be considered as school evenings, which, however, I at times do, though unwillingly, break through; and Tuesday and Friday I find little more than sufficient to do my own business in. So that you will perceive I have not much time to spare. Business is the first thing, to which I am not only tied by necessity, but by honour: pleasure is the last. And then, again, there is an intermediate part, verging on both, to which I consider it a duty to attend—I mean the Society. After my work, I attend to that, then to my own affairs, and then to my friends. This long explanation, however, looks so serious that I would cut it shorter if I had time, for I am confident it must be unnecessary; but being here, as it helps to fill up, I will leave it. I shall hope to see you on Friday; if not, you will let me know, and when you will come.

'I am, as always, yours sincerely,

'M. FARADAY.

'Excuse the haste.'

FARADAY TO ABBOTT.

'December 31, 1816.

'Dear A——, I have delayed writing for some days that I might, when I did do it, produce something of importance, in size at least, and I am in hopes from your last that you will not object to my intentions and reasons: the latter are, that our mutilated correspondence may be resumed to the advantage of us both. The observation contained in yours of the 25th, respecting the various causes and influences which have retarded our mutual communications, together with my own experience, which on this point you are aware has been great, make it desirable that our plans should be such as to facilitate the object we have in view in the writing of letters. That object is, I believe, the communication of information between us, and the habit of arranging in a proper and orderly manner our ideas on any given or casual subject, so that they may be placed with credit and service to ourselves on paper.

'This object, it strikes me, would in part be attained by giving, in addition to the general tone of a liberal and friendly letter, something of the essayical to our communications. I do not mean that every letter should be an essay, but that when a thought, or a series of thoughts, on any particular subject, for the moment enters the mind, that the liberty be allowed of throwing them into form on paper, though perhaps unconnected with what has gone before or may succeed it. This indeed, I believe, is the plan we have actually followed, but I am not sure that I had so conceived the thing before; at least I had not marked out in my own mind that in pursuance of our object I might sit down and scribble to you, without preface, whatever was uppermost in my mind. However, at present you perceive what I am at, and what are my intentions, and though, as I before observed, we have both virtually followed this plan, yet I have at this time given it something like form, or sound, or expression, or whatever else you please, that I might be more conscious of it, and make use to a greater extent of the liberty it allows me.

'I must confess that I have always found myself unable to arrange a subject as I go on, as I perceive many others apparently do. Thus, I could not begin a letter to you on the best methods of renovating our correspondence, and, proceeding regularly with my subject, consider each part in order, and finish, by a proper conclusion, my paper and matter together.

'I always find myself obliged, if my argument is of the least importance, to draw up a plan of it on paper, and fill in the parts by recalling them to mind, either by association or otherwise; and this done, I have a series of major and minor heads in order, and from these I work out my matter. Now this method, unfortunately, though it will do very well for the mere purpose of arrangement and so forth, yet it introduces a dryness and stiffness into the style of the piece composed by it; for the parts come together like bricks, one flat on the other, and though they may fit, yet they have the appearance of too much regularity; and it is my wish, if possible, to become acquainted with a method by which I may write my exercise in a more natural and easy progression. I would, if possible, imitate a tree in its progression from roots to a trunk, to branches, twigs, and leaves, where every alteration is made with so much care and yet effect, that though the manner is constantly varied, the effect is precise and determined.

'Now, in this situation I apply to you for assistance. I want to know what method, or what particular practice or exercise in composition, you would recommend to prevent the orderly arrangement of A_1 A_2 A_3 B_1 B_2

C_1 C_2 C_3 C_4, &c., or rather, to prevent the orderly arrangement from appearing too artificial. I am in want of all those conjunctions of styles, those corollaries, &c., by which parts of a subject are put together with so much ease, and which produce so advantageous an effect; and as you have frequently, in your contributions to our portfolios, given me cause to admire your success and lament my own deficiency upon this point, I beg that you will communicate to me your method of composing; or, if it is done spontaneously and without an effort on your part, that you will analyse your mental proceedings whilst writing a letter, and give me an account of that part which you conceive conducive to so good an end. With this request I shall refer the subject to you, and proceed briefly to notice the contents of your letter.

'With respect to my remarks on lectures, I perceive that I am but a mere tyro in the art, and therefore you must be satisfied with what you have, or expect at some future time a recapitulation, or rather revision, of them: but your observations will be very acceptable.

'M. FARADAY.'

The knowledge that remains of Faraday in 1817, like that of the previous year, is derived from four sources: his lectures, his commonplace-book, his publications, and his letters.

I.

At the City Philosophical Society he gave a lecture on the means of obtaining knowledge, and five chemical lectures: on the atmosphere, on sulphur, and phosphorus, on carbon, on combustion and on the metals generally.

In his commonplace-book there are many subjects marked for future work. There are also extracts from books, and some geological notes of a visit to Somersetshire and Devonshire, where he went to stay with his friend Huxtable, near South Moulton, for a month.

He had six papers and notices in the 'Quarterly Journal of Science.' The most important of these was an account of some experiments on the escape of gases through capillary tubes.

His letters to his friend Abbott show that the occupation of his time obliged him to write more shortly and much less frequently, but still it is from these letters that the best knowledge of his thoughts and acts can at this time be obtained.

Huxtable kept scarcely any notes of Faraday's visit to him. He says, 'In company with Mr. Flaxman, surgeon, and my brother Robert, we visited the copper mines in the neighbourhood of South Moulton. Mr. Flaxman gave Mr. Faraday a specimen of the gold we found at a mine in the neighbourhood. An account of this he published in the "Journal of the Royal Institution." Mr. Faraday experimented with Mr. Flaxman's imperfect galvanic apparatus.

'At the Narracote sheep-shearing, Mr. Faraday took part in the conviviality of the evening with much apparent interest and good-humour. I have had several opportunities of observing his social character since.'

The lecture which he read to the body of members at 53 Dorset Street, Salisbury Square, February 19, was, 'On some observations on the means of obtaining knowledge, and on the facilities afforded by the con-

1817. ÆT.25–26.

stitution of the City Philosophical Society.' This was printed by Effingham Wilson, 1817. It gives an account of the use he made of the Society as a means of education.

. . . . . .

'I have ventured,' he says, 'to bring this subject forward, under the disadvantage of having but a slight acquaintance with authors who have considered the matter, and under the still greater inconvenience of having but slightly considered it myself. My store of learning respecting knowledge, abstractedly considered, has been gathered, some time since, from the writings of Lord Bacon, and from a work by Dr. Watts, "On the Improvement of the Mind," which I consider so good in its kind that no person ought to be without it.

. . . . . .

'And here I must be allowed to say, that it is my firm belief, that were all the benefits which may be derived from a vigorous exercise and enjoyment of the powers and privileges of the City Philosophical Society well known and duly appreciated, each member would feel eager to share in the general good they present, and regret that such estimable advantages had been until now suffered to remain unemployed. For myself, I have perceived and used them; and it is but natural that one who has gained much by the Society should feel grateful for it, and endeavour to express it in terms of praise and respect. It has increased my store of mental enjoyment, and as it has taught me liberality I recommend it liberally to others. Nor can I refrain from saying that I know no institution, with means so small and professions so humble, calculated to produce so much effect or results so highly valuable.

'I trust I shall be excused for the warmth of my feelings on this occasion. I do not express myself thus because I imagine you are not conscious of the true value of the Society; but having experienced to a great extent its beneficial effects, I am willing to testify my consciousness of them. I shall now consider the means of obtaining knowledge:—

'*By Conversation.*

'It is with regret I observe that so little are our private evenings for conversation appreciated, that not one half of our members generally attend.

. . . . . .

'For my own part, so highly do I value the opportunities of our conversations, that I would rather be absent on any lecture night than on a private evening.

'*By Lecturing.*

'As practised at this Society, lecturing is capable of improving not only those who are lectured, but also the lecturer. He makes it, or he ought to make it, an opportunity for the exertion of his mental powers, that so by using he may strengthen them; and if he is truly in earnest, he will do as much good to himself as to his audience.

. . . . . .

'*By Reading.*

'With our parsimonious and economical subscriptions, it can scarcely be imagined that to apparatus, lectures, and conversations, a library should be added.

. . . . . .

'*By Observation.*

'There is one peculiar brancn of it for which we possess facilities as a body which many of us do not as individuals : I allude to the making of experiment.

. . . . . .

'*By Study.*

'Our Society is not at all deficient in those means which encourage a disposition to study : I refer more particularly to the portfolio, which has been established for the reception of such papers, analyses, or essays, either on lectures, questions, or independent subjects, as may be contributed by the members. This portfolio supplies the place of a report, and each one who comes forward is asked to place in it either his question or lecture; or, if he please, any original paper, so that it *should really* be the archives of the Society. It circulates among the members with the books of the library. . . I will not detain you longer, gentlemen, from the expressions of your opinions on this subject, than to point out to you two modes in which you may treat it. The question may be formally put, Whether the means of acquiring knowledge which I have pointed out are sufficient to the extent that I have described? Or, as I should rather wish it, the conversation may turn on the means afforded by the organisation of the Society, and on such improvements of those means as may suggest themselves to the members as being practicable.'

In the tenth lecture of his chemical course at the Philosophical Society, On Carbon, given July 16th, he began to use notes. In his previous lectures he had written out all that he intended to say. A few of the notes regarding the Safety Lamp are worth preserving on account of their reference to Sir H. Davy.

'The great desideratum of a lamp to afford light with safety: several devised; not mention them all, but merely refer to that which alone has been found efficacious, the DAVY: this the result of pure experimental deduction. It originated in no accident, nor was it forwarded by any, but was the consequence of a regular scientific investigation.

'The contest on lamps, disgraceful subject. Pass it over, except by noticing that those who have invented lamps are the most clamorous against the Davy.

'Eulogium.—An instance for Bacon's spirit to behold. Every philosopher must view it as a mark of subjection set by science in the strongest holds of nature.'

II.

In his commonplace-book the following entries show his thoughts at this time. Some of his questions he afterwards marked as answered:—

'On Mr. Hume's tests for arsenic: changes produced in the colour of bodies by heat alone.

'*July.*—Geological notes: South Moulton slate; Tiverton; Hulverston; leave Devonshire, Taunton, Somerton, Castle Cary.

'*Chemical Queries.*

'The action of oxide, chloride, fluoride, and iodide of silver on ammonia, and the nature of the compound formed?

'The substance formed by fused muriate of lime and ammoniacal gas; its nature?

'Sulphuret of phosphorus; its true proportions and properties? Chromic Compounds, particularly those with base of chrome; also chromate of chrome, and the analysis?

'Exciting effects of different vapours and gaseous mixtures?

'Silvering of silk and other animal substances by ammonuret of silver? Phosphuret of carbon, its nature the green powder? Black dye for silk. Combination of ammonia with chlorides. Arsenic acid as a test to discriminate between barytes and strontia, and also as a test for zinc.

'Muriate of silver and ammonia. Melting of horn. Tellurium on sulphur. Chlorine and carbon, made out autumn of 1820.

'Mutual action of muriate and nitrate of ammonia.

'Electricity. Magnetism. A pyrometer. Tests for barytes, strontia, and lime; made out.

'What is the acid which is generated in stale infusion of red cabbage? Query, the nature of the change of colour induced by sub-borate of soda in solution on cabbage infusion? It does not render it green, though strongly alkaline.

'Passages out of "Shakespeare," "Lalla Rookh," "Johnson," "Morning Chronicle;" "Chemistry a Corrective of Pride," Klaproth's "Analyses of the Blood of a German Noble and his Servant, identical."

'Reference to "Rambler," on uncertainty of life; secresy; fancied virtue; prescience; pertinacity of opinion; forgiveness of injuries; friendship; moral virtue; good-nature; peevishness; on sanguinary laws; on literary courage; impropriety of haste in life.'

III.

The papers which he published in the second and third volumes of the 'Quarterly Journal of Science' show the researches on which he was at this time engaged: 1. Some account of the alstenia teiformis, or tea of Bogotá. 2. Report of some experiments made with compressed oxygen and hydrogen, in the laboratory of the Royal Institution. 3. Notice of some experiments on flame made by Sir H. Davy. 4. On the wire-gauze safety lamps. 5. Some experiments and observations on a new acid substance formed from ether. 6. An account of some experiments on the escape of gases through capillary tubes.

IV.

Among the letters which he wrote this year, that to his friend Abbott, upon the death of his mother, shows the kindness of his nature. In other letters, his occupations and general advance in science are well seen.

FARADAY TO ABBOTT.

'Royal Institution: January 20, 1817.

'Dear A——, The irresistible propensity in the human breast to draw conclusions, before every circumstance has been examined, or even before possession has been obtained of the necessary data, is so general, that it passes unnoticed, although constantly active in ourselves, until some very flagrant instance in others draws the attention to the results of such irregular and improper proceedings, and points out the folly of immature judgments. Now, though it happens that these flagrant instances occur frequently, and are continually reminding us of our delinquency, yet, somehow or other, the fault still retains its ground, and even appears at times to increase in strength.

''Twould be a source of much useful consideration to endeavour to point out those causes which support and strengthen this ill habit of the mind, and the comparative strength of them in persons differing in intellectual powers and tempers. It is not, however, my intention at this time to take up this subject, though I am conscious it would* be of much service to me, by giving me a more direct and positive knowledge of this effect; but I have been led thus far into the subject by hearing from Farley, that Mr. Murray had informed him Sir H. Davy had stolen some experiments from the French chemists, and adopted them as his own.

'(I received yours of the 17th at the above,* but, being determined to finish my sentence, proceeded. To continue)—

'Murray has told Farley that Sir Humphry keeps a platinum wire red-hot by holding it over ether, and he says the effect is produced by the sulphur of the ether combining with the platinum, and that the experiment is the same with that of the French chemists, where they combine sulphur and copper-leaf directly. Now, lest you should be troubled by the queer explanation of an impossible effect, I shall (being now permitted) lay this (a) new discovery before you.

'Sir Humphry has lately been engaged in an investigation on the nature of flame . . .

. . . . . .

making experiments with platinum in mixed oxygen and hydrogen, in coal-gas, in ether vapour, and hot alcohol.

'These experiments succeed with all the combustible gases and vapours, even with that of warm alcohol, but the only metals that are efficacious are platinum and palladium; the others possess too much radiating and conducting power. I need not point out to you the inapplicability of Murray's explanations. They will strike you at once, and perhaps to-morrow evening, should he accost you, you will be able to inform him on *this* point as well as on many others.

. . . . . .

'What with Sir Humphry, Mr. Brande, our two-fold series of lectures, original investigations, the Society and its committee, my time is just now so closely cut up, that Sunday will hardly suffice for my mother, brother, and sister; and as your hospitality constantly presses me to dinner, which when accepted as constantly makes me too late, I hardly know what to do. I have determined, as far as I can, to see you next Sunday, but write me for fear of a failure.

. . . . . .

'I am, dear B., yours ever,

'M. FARADAY.

'I never can get through a letter with any regularity, and so you must excuse everything.'

FARADAY TO ABBOTT.

'Royal Institution: June 9, 1817.

'My dear B——, When are you going, or rather coming, to ramble my way? Can it be on next Friday evening, for on that evening I shall have the privilege of doing my work in my own room, and can of course talk to a friend? I now begin to get sight of a period to this busy time, and hope by the beginning of next month to have cleared away the mass of clearings, and preparations, and arrangements, &c., that now impede my way. Our lectures are nearly over. Evening meetings will soon cease for the season, and then I really mean to relax.

'Come, if you can, on Friday, till when good-bye.

'Yours,

'M. FARADAY.'

FARADAY TO HIS MOTHER.

'Barnstaple: June 27, 1817.

'Dear Mother,—I seize a spare moment to write you from this place, where we arrived a few hours since, having had a pleasant walk and ride from South Moulton.

'I am sure that the interest you feel in me will make you desirous of knowing how my health stands, and I have much pleasure in telling you that in that respect I am improved in every way. My strength is greatly increased, all my scars have disappeared; I am growing quite merry, and am in every way far superior to what I was.

'I trust that you also are well, with M., and all the rest of our friends; and further, I hope I shall find the affairs of the house well and comfortably arranged.

'I have seen a great deal of country life since I left town, and am highly pleased with it, though I should by no means be contented to live away from town. I have been at sheep-shearing, merry-making, junketings, &c., and was never more merry; and I must say of the country people (of Devonshire, at least) that they are the most hospitable I could imagine. I have seen all your processes of thrashing, winnowing, cheese and butter making, and think I could even now give *you* some instructions; but all I have to say to you on these subjects shall be said verbally.

'We are just now moving a little about the country, and I find myself much interested by what I meet with. It would not, however, afford you the same pleasure, for to talk of wavellite, hydrargellite, and such hard things, would be out of the question when the question was to you.

. . . . . .

'I am, dear Mother, your affectionate son,

'M. FARADAY.'

FARADAY TO ABBOTT.

'Royal Institution: Friday morning,
half-past seven, July 25, 1817.

'Dear A——, I did not get your kind note until last night, and since then have been endeavouring to extricate myself from a meeting of some of our people, respecting a singing school attached to our meeting-house, in hopes that I could have said to you, "I will come." It happens, however, unfortunately, that I have been one of the most earnest in bringing the consultants together, and that I shall be obliged to be with them, plodding over the means of improving our own singing, in place of attending to and enjoying that of others, on this evening.

'It avails one nothing to express my regrets; therefore I shall restrain my pen, except in saying I am, as always, yours sincerely,

'M. FARADAY.'

FARADAY TO ABBOTT.

'Royal Institution: November 25, 1817.

'Dear A——, I can scarcely imagine the opinion you will form of me in seeing me thus changeable; but this, however, I think I may safely presume, that you will not charge me with ceremony. I do not know what the fates intend by interfering with our arrangements, but I know to my cost how they do, and I have now to let you know. This is to inform you that, in consequence of an arrangement I have made with a gentleman recommended to me by Sir H. Davy, I am engaged to give him lessons in mineralogy and chemistry thrice a week in the evening, for a few months. In order to meet the engagement, I am obliged to neglect my Monday evening school entirely, and to give up with that my Tuesday and Thursday evenings to teaching. Our lessons do not commence till eight o'clock, and, as my gentleman is in the immediate neighbourhood, I am at liberty till that hour. I hope, therefore, that I shall see you early on Thursday, but I regret that I must also part with *you* early. We will look over the letters, &c., as you desire.

'Yours sincerely,

'M. FARADAY.'

Faraday's progress in knowledge in 1818 can still be

well marked, (1) by his lectures; (2) his note-book; (3) his publications; and (4) his letters.

Continuing his course of chemical lectures, he gave five this year: these were on gold, silver, mercury, &c.; on copper and iron; on tin, lead, zinc; on antimony, arsenic; on alkalies and earths.

He also gave another lecture, entitled 'Observations on the Inertia of the Mind.' This lecture was written out at full length: some passages, which show Faraday's own mind at this period, are worth preserving.

His note-book contains a carefully reported course of lectures on oratory, by Mr. B. H. Smart, who still can tell of the strong desire of his pupil to improve himself, although fifty years have passed since Faraday, with little time and less money, thought it well to do all that could be done for his own education even in his manner of lecturing.

He had eleven papers in the fourth and fifth volumes of the 'Quarterly Journal of Science.' The most important was on sounds produced by flames in tubes.

As Faraday's scientific work increased, his letters to his friend Abbott became fewer and shorter. This year a very short one almost ends the correspondence. As long as Abbott stayed in London, the friends met from time to time; and in after years, when he chanced to be in town, his greatest pleasure was to witness the success of Faraday in the theatre of the Royal Institution.

The loss of the reflection of Faraday's mind in these letters is partly balanced by another image of himself, which becomes visible in a correspondence which he began this year with Professor G. de la Rive.

When Faraday was at Geneva with Davy, Professor

de la Rive, undazzled by the brilliancy of Davy's reputation, was able to see the true worth of his assistant. This led him to place Faraday, in one respect, on an equality with Davy. Whilst they were staying in his house, he wished them to dine together at his table. Davy, it is said, declined, because Faraday acted in some things as his servant. De la Rive expressed his feelings strongly, and ordered dinner in a separate room for Faraday.

Of that Geneva visit Faraday says, in 1858, to Mr. A. de la Rive, 'I have some such thoughts (of gratitude) even as regards your own father, who was, I may say, the first who personally at Geneva, and afterwards by correspondence, encouraged and by that sustained me.'

It will be seen that this correspondence, which began with the father, was continued with the son, and it lasted altogether nearly fifty years; and there was no one to whom Faraday wrote an account of his work and of his thoughts with so much pleasure and so much sympathy as to his friend Professor A. de la Rive.

I.

In 1818 his lectures still give the best insight into his mind.

At the end of his thirteenth lecture, on gold, silver, mercury, platinum, palladium, rhodium, iridium, osmium, he says:—

'As in their physical properties so in their chemical properties. Their affinities being weaker, (the noble metals) do not present that variety of combinations, belonging to the more common metals, which renders them so extensively useful in the arts; nor are they, in consequence, so necessary and important in the operations of nature. They do not assist in her hands in breaking down rocks and strata into soil, nor do they help man to make that soil productive or to collect for him its products.

'The wise man, however, will avoid partial views of things. He will not, with the miser, look to gold and silver as the only blessings of life; nor will he, with the cynic, snarl at mankind for preferring them to copper and iron. He will contemplate society as the proper state of man, and its artificial but necessary institutions and principles will appear to him the correct and advantageous result of natural causes. That which is convenient is that which is useful, and that which is useful is that which is valuable. It is in the relative position of things one to the other that they are to be considered and estimated; and whilst a man makes use of them no otherwise than wisely to supply his wants and virtuous pleasures, the avaricious trader has no reason to call him a fool of nature, nor the moral philosopher to name him the victim of society.'

His fourteenth and fifteenth lectures were on copper, iron, tin, lead, and zinc.

At the end of the sixteenth lecture, on antimony, arsenic, cobalt, manganese, nickel, bismuth, tungsten, molybdenum, uranium, cerium, tellurium, titanium, columbium, he says:—

'I have now, in the progress of the three last lectures, brought before you by far the most important of the metallic bodies. There yet remain a few, the result of late, even of contemporaneous research, which will form the subject of the next and the concluding lecture of this course. They are the bases of the alkalies and alkaline earths. It is interesting to observe the progress of this branch of chemistry, its relations to the ages through which it has passed, and the continual refinement of the means which have urged it in its career.

'The ancients knew but of seven reduced metals, but they are those which, amongst the extensive range we now possess, are the most important; and in gold, silver, mercury, copper, iron, tin, and lead they found abundant resources for weapons of war and for implements of art, for economical applications and for ornamental uses.

'The metals in use in old times were obtained almost by accident—either pure from the hands of nature or by the rudest and simplest workings. But, excited by the result of their labours, and by a rude perception of the important ends to be obtained, men would, in the course of years, not only have their curiosity, but their interest engaged in the pursuit; and, improved by the experience of past ages, we find the alchemists and their followers, in spite of the self-created obscurity which surrounded them, still frequently successful in withdrawing from the concealed stores of nature new metallic wonders, and giving to mankind at one time amusing, at another useful information.

'As the views of men became clearer, as their growing means continually improved by practice in their hands, new individuals were added to the metallic species, and each addition drew forth applause for the genius of the discoverer, and for his contribution to general chemical science. Stimulated by the due and awarded commendation given to prior merit, all exerted themselves; and the result at this day is, that in place of the seven metals known to the ancients, at least forty have been distinguished from each other and from other bodies, and have had their properties demonstrated.

'At present we begin to feel impatient, and to wish for a new state of chemical elements. For a time the desire was to add to the metals, now we wish to diminish their number. They increase upon us continually, and threaten to enclose within their ranks the bounds of our fair fields of chemical science. The rocks of the mountain and the soil of the plain, the sands of the sea and the salts that are in it, have given way to the powers we have been able to apply to them, but only to be replaced by metals. Even the gas of our atmosphere puts on at times a metallic appearance before us, and seems to indicate a similar base within. But a few combustibles and a few supporters of combustion remain to us of a different nature, and some (men of celebrity too) have found metals even among these.

'To decompose the metals, then, to reform them, to change them from one to another, and to realise the once absurd notion of transmutation, are the problems now given to the chemist for solution. Let none start at the difficult task, and think the means far beyond him; everything may be gained by energy and perseverance. Let us but look to the means which have given us these bodies, and to their gradual development, and we shall then gain confidence to hope for new and effective powers for their removal from the elementary ranks. Observe the first rudiments of metallurgical knowledge in the mere mechanical separation of native gold and silver from the encumbering substances; mark the important step made in the reduction of copper, of iron, and of tin from their ores—rude, indeed, in the hands of the ancients, but refinement itself compared with their prior knowledge; consider the improvement when, by a variety of manipulations, the early chemist of the last century separated a small quantity of a metallic substance from five or six other bodies, where it existed in strong combination, and then pass to the perfection of these means as exhibited in the admirable researches of Tennant and Wollaston; lastly, glance but at the new, the extraordinary powers which the chemist of our own nation put in action so successfully for the reduction of the alkalies and earths, and you will then no longer doubt that powers still progressive and advanced may exist and put at some favourable moment the bases of the metals in our hands.'

At the conclusion of his course, at the end of the seventeenth lecture, on alkalies and earths, he says:—

'The substances that have now been described complete the series of metallic bodies which I enumerated to you among the present elements of chemical science. With them we conclude our consideration of simple bodies, and at this point too it is my intention to close the present series of lectures.

'During the seventeen lectures which have formed this course, I have constantly endeavoured rather to enlarge upon the powers and properties of the simple substances, and upon a few of their more proximate combinations, than to describe to you their ultimate compounds, and their applications and subserviency to the purposes of man. I have consequently drawn my matter from more hidden and obscure sources, of the character of which it has partaken, and have, to a certain degree, neglected to use that influence over your attention which would have been in my power if I had reverted to effects frequently in common life before you, and to bodies the uses of which are daily in your hands.

'It certainly is not necessary that I should give reasons for the adoption of a particular plan, though it may appear to be imperfect and badly designed; but courtesy claims some attention even in science; and, influenced by this, and in justification of myself (for whatever independence we may assume, no one is inclined to permit a censure, however unjust or however slight, to rest upon him which he can remove), I *shall* give reasons.

'On considering at the commencement of the course what ought to be the nature of these lectures, they appeared to me to admit with propriety of two distinct characters: they might be illustrative of the various processes and applications in the arts which are of a chemical nature; or they might be elementary in their nature, and explanatory of the secret *laws and forces* on which the science, with all its uses, is founded.

'These two modes in which chemistry may be said to exist, you will observe, form what may be called the extremes of our knowledge on the subject; the first, or applicative chemistry, is identical with the experimental knowledge and practice of the artisan and manufacturer, and is also that from which we gain our first perceptions and ideas of this kind; the second, or elementary chemistry, is the result of our researches *in the science*; and though, being the result of long and laborious trains of inductive reasoning and experimental investigation, it is the last production of the mind, yet it is also the basis upon which nature and art have raised practical science.

'To have made these lectures, therefore, illustrative of the arts and manufactures, would have been to repeat what the man of observation had already noticed, i.e. the results of general experience: and as it cannot but be supposed that a knot of men drawn together for mutual improvement must consist of men of observation, it would have been to repeat at least much of what must be known to the members of the Society. On the contrary, to treat of the first laws and principles of the science would be to explain the causes of the effects observed by those attentive bystanders on nature, and to point out to them by what powers and in what manner those effects were produced.

'It is possible, and likely, that to suppose men not observant would be to incur their distaste and enmity, but to suppose them ignorant of elementary chemistry (and by lecturing I have supposed, in form at least, ignorance) could be no cause of offence. The acute man may observe very accurately, but it does not follow that he should also reason perfectly or extensively. To tell a person that a stone falls to the ground would be to insult him, but it would not at all compromise his character for sagacity to be informed of the laws by which the stone descends, or the power which influences it. And though perhaps this illustration does not apply *very strongly* to the case of *applicative* and elementary chemical science, it is still satisfactory, by assuring me that I have not given offence by any imaginary depreciation of the knowledge of others. But, further, I know of no illustration of the arts and manufactures of the civilised world, which would have been worth my offering, or your acceptance, not founded on first principles. I have endeavoured to give those first principles, in an account of the inherent powers of matter, of the forms in which matter exists, and of simple elementary substances. It is probable that at a future time I may make use of this groundwork in illustrating the chemical arts to you; in the meantime I shall rest confident that, should I find it convenient, the sufferance and kindness you *have* extended to me will still attend my efforts.

'Before I leave you, I may observe that, during the last year or two of the Society's proceedings, I have been intruded on your attention to an extraordinary degree: this has been occasioned partly by my eagerness to use the opportunities this Society affords for improvement; partly by the necessity we are under, from the constitution of our laws, to supply every required effort of whatever nature from among ourselves; but much more by *your* continued indulgence.

'At present I shall retire awhile from the public duties of the Society, and that with the greater readiness, as I have but little doubt that the evenings devoted to lectures will be filled in a manner more worthy of your attention. I retire gratified by the considerations that every lecture has tended to draw closer the ties of friendship and good feeling between the members of the Society and myself; that each one of them has shown the advantages and uses of the Society; and still more, by the consciousness that I have endeavoured to do, and the belief that I have done, my duty to the Society, to myself, and to you.'

In his 'Observations on the Inertia of the Mind,' read at the City Philosophical Society July 1, 1818, he said: 'It is now about eighteen months ago that I read in this room a few observations on the means of obtaining knowledge, and on those which particularly interested us as being afforded by the principles and constitution of this Society. My object at that time was to describe distinctly some of the common and accessible sources of information, with the means best adapted to gain for each of us a participation in the benefits and the honours presented by knowledge to mankind. In continuance of that subject, it is my intention to-night to attempt an exposition of certain influences which retard, if they do not prevent, the application of those means, and which oppose the appropriation to ourselves of that to which we have an inalienable right.

. . . .

'Man is an improving animal.

'Unlike the animated world around him, which remains in the same constant state, he is continually varying; and it is one of the noblest prerogatives of his nature, that in the highest of earthly distinctions he has the power of raising and exalting himself continually. The transitory state of man has been held up to him as a memento of his weakness: to man *degraded* it may be so with justice; to man as he ought to be it is no reproach; and in knowledge, that man only is to be contemned and despised who is *not* in a state of transition.

'We are by our nature progressive.

'We are placed by our Creator in a certain state of things, resulting from the pre-existence of society, combined with the laws of nature. Here we commence our existence, our earthly career. The extent before us is long, and he who reaches furthest in his time has best done his duty, and has most honour. The goal before us is perfection: always within sight, but too far distant to be reached. Like a point in the utmost verge of perspective, it seems to recede before us, and we find as we advance that the distance far surpasses our conception of it. Still, however, we are not deceived; each step we move repays abundantly the exertion made, and the more eager our race the more novelties and pleasure we obtain.

'Some there are who, on this plain of human life, content themselves with that which their predecessors put into their possession, and they remain idle and inactive on the spot where nature has dropped them; others exist who can well enjoy the advantages in advance, but are too idle to exert themselves for their possession—and these are well punished for the envy which their very sensibility and sentient powers engendered within them at sight of the success of others; a third set are able and willing to advance in knowledge, but they must be led; and but few attain to the distinguished honour of being first on the plain, and of taking the lead of their generation, of the age, and of the world.

'It can scarcely be possible that my opinion should be mistaken in what I have said; but, lest anyone misconceives me, I shall take the liberty of discriminating some few points before I proceed further in asserting the improvability of man.

'First, then, all theological considerations are banished from the Society, and of course from my remarks; and whatever I may say has no reference to a future state, or to the means which are to be adopted in this world in anticipation of it. Next, I have no intention of substituting anything for religion, but I wish to take that part of human nature which is independent of it. Morality, philosophy, commerce, the various institutions and habits of society, are independent of religion, and may exist either with or without it. They are always the same, and can dwell alike in the breasts of those who from opinion are entirely opposed in the set of principles they include in the term religion, or in those who have none.

'To discriminate more closely, if possible, I will observe that we have *no* right to judge religious opinions, but the human nature of this evening is that part of man which we *have* a right to judge; and I think it will be found, on examination, that this humanity—as it may perhaps be called—will accord with what I have before described as being in our own hands so improvable and perfectable.

'Lastly, by advancement on the plain of life, I mean advancement in those things which distinguish man from beasts—sentient advancement. It is not he who has soared above his fellow-creatures in power, it is not he who can command most readily the pampering couch or the costly luxury; but it is he who has done most good to his fellows, he who has directed them in the doubtful moment, strengthened them in the weak moment, aided them in the moment of necessity, and enlightened them in their ignorance, that leads the ranks of mankind.

'Such then is our state, and such our duty. We are placed in a certain point in the immensity of time, with the long, the interminable chains of moral good and of human knowledge lying about our path. We are able to place them straight before us, to take them as our guides, and even to develope them to others far beyond the spot where we found them; and it is our duty to do so. Some there are perverse enough to entangle them even wilfully, to delight in destroying the arrangements which nature points out, and to retard their very neighbours in their efforts; but by far the greater part are content to let things remain as they are. They make no efforts themselves, but, on the contrary, hang as weights upon the exertions made by others who labour for the public good. Now, it is with the spirit which animates, or rather benumbs, these that I would have to do. I trust there are not many who retrograde, and for the sake of human nature I will not believe that the observations which apply to them should be general.

'There is a power in natural philosophy, of an influence universal, and yet withal so obscure, in its nature so unobtrusive, that for many ages no idea of it existed. It is called *inertia.* It tends to retain every body in its *present* state, and seems like the spirit of constancy impressed upon matter. Whatever is in motion is by it retained in motion, and whatever is at rest remains at rest under its sway. It opposes every *new* influence, strengthens every *old* one. Is there nothing in the human mind which seems analogous to this power? Is there no spiritual effect comparable to this corporeal one? What are habits? old prejudices? They seem something like a retention in a certain state due to somewhat more than the active impulses of the moment. As far as regards them, the mind seems inclined to remain in the state in which it is, and the words which enunciate part of our natural law will describe exactly the effect. The agreement is strange, but it nevertheless is evident and exact thus far; and it is possible we shall find it to exist even in its more active states. We have only to ascertain whether the mind which has once received an impulse, which has become active and been made progressive, continues in that state, and we can decide at once on the analogy. I do not know whether you will require of me to prove that such is the case before you will admit it. The impression on my own mind is that it is eminently so; and I doubt not but that your own observation will confirm my conclusions. The man who has once turned his mind to an art goes on more and more improving in it; the man who once begins to observe rapidly improves in the faculty. And to illustrate at once the force of mental inertia to retain the mind either at rest or in motion, how difficult our endeavours to set about a new affair, how facile our progress when once engaged: every little delay illustrates more or less the inertia of the passive mind; every new observation, every fresh discovery, that of the active mind.

'Perhaps in playfulness we may endeavour to trace the analogy still further. Inertia is an essential property of matter; is it a never-failing attendant on the mind? I hope it is; for as it seems to be in full force whenever the mind is passive, I trust it is also in power when she is actively engaged. Was the idle mind ever yet easy to be placed in activity? Was the dolt ever willing to resign inanity for perception? Or are they not always found contented to remain as if they were satisfied with their situation? They are like the shepherd Magnus: although on a barren rock, their efforts to remove are irksome and unpleasant; and they seem chained to the spot by a power over which they have no control, of which they have no perception. Again: in activity, what intellectual being would resign his employment? Who would be content to forego the pleasures hourly crowding upon him? Each new thought, perception, or judgment is a sufficient reward in itself for his past labours, and all the future is pure enjoyment. There is a labour in thought, but none who have once engaged in it would willingly resign it. Intermissions I speak not of: 'tis the general habit and tenor of the mind that concerns us, and that which has once been made to taste the pleasures of its own voluntary exertions will not by a slight cause be made to forego them.

'There is still another point of analogy between the inertia of matter and that of the mind, and though not essential in either case, yet the circumstances exist in both. I refer to what may be called disturbing forces. If the inertia of matter were to be exerted alone, it would tend, according to the original state of rest or of motion, to preserve the universe eternally the same, or to make it ever changing. At present it is doubtful whether both these effects do not take place, but they certainly do not happen in the same manner. The centripetal force, the centrifugal force, the force resulting from chemical action, and that which originates muscular exertion, are at all times active in changing and varying the states induced by inertia, sometimes aiding, sometimes counteracting its effect. These are represented among intellectual beings by the sensations, perceptions, passions, and other mental influences which interfere (frequently so much to our inconvenience) with the dictates of our reason. The philosopher who has perceived and enjoyed the advantages resulting from the actual performance of his own experiments and the use of his own senses, has all his industry (I would say inertia) destroyed by the lassitude of a hot day, and gravitates into inactivity. Another has his reasoning crossed by his inclination; some thoughts are driven one way, some another, and his mind becomes a mere chaos. Others there are, again, whose inertia is assisted by the repeated action of other causes, and they go on with accelerated energy. So vanity, ambition, pride, interest, and a thousand other influences, tend to make men redouble their efforts, and the effect is such that what appeared at first an impassable barrier easily gives way before the increasing power opposed to it.

'Inertia, as it regards matter, is a term sufficiently well understood both in a state of rest and of motion. As it is not my intention to attempt a description of functions of the mind according to strict mathematical terms, I shall resign the exclusive use of the word at present, and adopt two others, which, according to the sense they have acquired from usage, will, I believe, supply its place with accuracy. Apathy will represent the inertia of a passive mind; industry that of an active mind.

'It is curious to consider how we qualify ideas essentially the same, according to the words made use of to represent them. I might talk of mental inertia for a long time without attaching either blame or praise to it—without the chance even of doing so; but mention apathy and industry, and the mind simultaneously censures the one and commends the other. Yet the things are the same: both idleness and industry are habits, and habits result from inertia.

'Let us first consider the inertia of the sluggish mind. This is apathy—idleness. Perhaps there never yet was a person who could be offered as a complete instance of this state—one who made not the slightest advancement in the paths of knowledge. It is not possible there should be such, where perception and reason exist though but in the slightest degree. Society must of necessity entrain such a being, even though against his will; and he will be moved like the rocky fragment in the mountain torrent—a fit display of the energetic powers about him, and of his own mean, inanimated state.

'But for want of this complete illustration, we may select instances where the general character is heavy and dull, or where that idleness which has been dignified unjustly with the name of *contentment* exists. Or we may take a particular individual, and select such parts of his character as are most subject to the benumbing influence of apathy. We cannot fail of finding (each in his circle) plenty like to the first of these; and perhaps I should not assert too much if I were to say we are all included among the latter.

'What, then (in the name of Improvement I ask it), what is the reason that, with all these facilities, without a single apparent difficulty, we are destitute in subject and meagre in interest? Alas! it must be apathy; that minister of ignorance has spread his wing over us, and we shrink into indolence. Our efforts are opposed by his power, and, aided by our mean sensations of ease, he triumphs over our better judgment, and thrusts it down to contempt. And is it possible that a being endowed with such high capabilities as man, and destined to such eminent purposes, should see his powers withered, his object unattained, through the influence of that mean thing, habit, and *still* remain content? Can it be that the *degradation* and a *consciousness* of it exist at one and the same time in the same being? or has apathy so powerful an agent in self-complacency that conviction is put to flight, and allowed no place in the breast? Whatever be the reason, the melancholy truth is evident, that we are fit for the noblest purposes, but that we fulfil them not.

'Nor is it over that appropriation of the reasoning powers alone which constitutes literary and scientific knowledge that this demon sways his withering sceptre. He triumphs also over the busy walks of commerce, and, alas! in the humble paths of morality. You shall sometimes see a tradesman set out in life with excellent prospects, stimulated by hope, ambition, interest, emulation, the incitement of his friends, and his own gratifications; he will exert every nerve to secure success, and he will succeed. You shall see this man gain on the world, till he stands, a fair example to others of the prosperity attendant on industry. He is raised above want, even the want of a luxury. But later in life you shall see this man 'fall from his high estate,' and, accompanied by repinement, regret, and contempt, sink into poverty and misery. The cause is idleness—apathy. Early in life he had been stimulated to personal exertions, and his due reward was prosperity. With it, however, came enjoyment, and as his wealth increased, so did the love of its pleasures. That full draughts might be taken of the sweetened cup of life, he resigned his cares into the hands of subordinate managers, and gave himself up to habits of enjoyment and ease. The strong interest which made his affairs prosper no longer governs them; but a secondary feeling actuates those to whom they are entrusted, and the energy of the measures taken for their preservation sinks in proportion. There is now no individuality between the results and the manager. Neglect creeps in; the shadows of confusion come over. At some thwarted moment the master *sees* this; he would fain rise to activity, but habit has imperceptibly taken possession of him. He struggles into exertion; but his exertions are momentary, and he falls again into supineness. Delays retard the aid he should bring, whilst they accelerate the fate attending him, until at last, when too late, the bright prospect and the solid realities recede together into obscurity and chaos.

'You will tell me perhaps that this is imaginary, or that at most it occurs but now and then, in very insulated instances; but I will give it as my own opinion that every tradesman realises it more or less. Where is the man who has used his *utmost* exertions in the prosecution of his affairs? Where is the person who has never relaxed but when fatigue required it? Has pleasure never taken place of business at an inconvenient moment? Has an appointment never been missed through careless delay? Has anyone reason to congratulate himself that he has lost nothing through inattention and neglect? If you assent to what these questions imply, you assent to my proposition, and allow that apathy is stronger at times even than interest.

'In morality I fear I should not have so difficult a task in establishing my assertion as in interest. When we continually see the former giving way under the influence of the latter, there is but little hope that it should withstand this influence, or that that which has conquered the stronger power shall not overcome the weaker. Morality seems the natural impression of the Deity within us. It ministers only to the serene and healthy but quiet pleasures of the heart, and has little to do with the passions and gratifications of the human being of this age. It is continually buffeted about in the tempest of temporal excitement, and rarely fails to suffer. Perhaps if the human being were placed out of the sphere of earthly influence, were not dependent upon it for support, and found no tempting pleasures in its productions, he might become conscious, even spontaneously, of the gratifications arising from the fulfilment of duties, and become more and more virtuous for virtue's sake; but, crossed as his good resolutions are by temptations and excitements, some effect must be produced which tends to warp the result of his conviction, and prevent that progression which ought to have place. Some take the system of morality as they find it for their standard, and act no further than it directs, forgetting that the institutions and the abuses of society frequently sanction vice of the most gross or the most contemptible kind. Ask the glutton whether moderation is a virtue and excess a vice. He will tell you yes, but his moderation is eating, drinking, and feasting; and his excess only that which produces dropsy, apoplexy, and death. Ask the gentleman what is the greatest disgrace to him; he tells you, a lie; but all the falsities of civilised and polite life are excluded from that term when *he* uses it. The morality of these persons, therefore, is the convenient system they have made up for themselves; gratification generally prevents them from perceiving any other, and if it fails, apathy secures them from any improvement. Others can perceive the right and the wrong, and have no objection to inculcate the purest principles. They do not like, however, to resign the pleasures they can secure by a slight practical trespass of their own rules; and that which in reality is the result of degraded taste and idleness they call expediency, and excuse their little derelictions from pure virtue by naming them necessary submissions to the present state of things.

'But leaving this melancholy picture of the effects of apathy on the human mind in your hands for consideration, I shall hasten to put an end to these observations by a few words on the effect of industry or the inertia of an active mind.

'Industry is the natural state of man, and the perfection of his nature is dependent on it; the progression which distinguishes him from everything else in the material world is maintained by it alone. The sun rises and sets, and rises and sets again. Spring, summer, autumn, and winter, succeed each other only to be succeeded by the same round. A plant rises from out of the earth, puts forth leaves and buds; it strengthens, arrives at maturity, and then dies, giving place to other individuals who traverse the same changes. An animal is born, grows up, and at last gives signs even of intelligence; but he dies without having improved his species: and it is *man alone* who leaves a memento behind him by his deeds of his having existed; who surpasses his predecessors, exalts his present generation, and supports those that follow him.

'Mere effects, however, which distinguish him from every other animated being, are only to be produced by industry; it is that which enables him to add to the

1818. ÆT. 26.

sum of knowledge already in possession of the world, to increase the stock of good which ennobles his nature. If he be not active, not in a state of improvement, what better is he than the brutes? In his own nature, none; and it is only what society has superinduced upon him of *its* manners and customs that distinguishes him from them.

'Dryden, I think, wrote an epitaph upon such an one, and it is very expressive of the vacuity of character and paucity of interest which such a being possesses or excites :—

'Here lies Sir John Guise : no one laughs, no one cries.
Where he's gone, how he fares, no one knows, no one cares.

'I have already endeavoured to establish the analogy between a habit of industry and the inertia of a moving body ; and as I fear I have too much trespassed upon your time and your good judgment in the foregoing attempts, I shall not further pursue it. I am sure that it is not needful for me to point out the good effects which would result to the Society from the *active* exertion of its members ; they must either have felt conscious of it already, or otherwise have found reasons against me which it would be politic in me first to hear. I shall, in conclusion, merely make an observation which I trust will extenuate me from the charge of harshness, and put a question (rather for form's sake than the question itself), with which I shall leave it in your hands.

'I have said that the inertia of matter is continually blended with other forces which complex its results and render them apparently contrary to their cause, and also that in this respect it resembles the inertia of the mind. This of course is equivalent to an avowal that there are *natural* disturbing forces of the inertia of the mind, and that an irregular, a retarded, or even an inverted progression must at times take place in knowledge and morality without any gross charge being incurred by mankind. I do not deny it. It was not, however, my particular object to discuss these forces, but the more general and fundamental one. If any, therefore, feels offended with what may appear like animadversions, he is at perfect liberty to take shelter behind these extenuations and secure himself from censure.

'In pursuing the analogy in my own mind of this general influence to which both matter and mind are subjected, I was led to a conclusion respecting mental inertia, which, though I have no reason to doubt, I should be fearful of uttering on my own authority alone. I will therefore put it in the form of a query; supposing, however, that still you will direct your conversation, if you feel incited, as much to the current remarks as to the question which will terminate them. Inertia has a sway as absolute in natural philosophy over moving bodies as over those at rest. It therefore does not retard motion or change, but is as frequently active in continuing that state as in opposing it. Now, is this the case with mental inertia?

'That I may ask the question more distinctly, I will preface it by two others, which, if disallowed, will give rise to conversation; if allowed, will prepare for the third. Are there not *more passive* than active minds in the world? Is mental inertia as puissant in active as in idle cases? Then, what is the cause of the state implied by the first question? or what is the reason why, unlike the material world, there is so much more of inanimation than of activity in the intellectual world?'

Such were Faraday's thoughts at this time, and from his commonplace-book more of his mind and work during this year can be seen:—

His notes begin with Lectures on Oratory, by Mr. B. H. Smart. These were very fully reported, filling one hundred and thirty-three pages.

Questions for Dorset Street. An experimental agitation of the question of electrical induction. 'Bodies do not act where they are not:' Query, is not the reverse of this true? Do not all bodies act where they are not, and do any of them act where they are? Query—the nature of courage, is it a quality or a habit?

Query—the nature of pleasure and pain, positive, comparative, and habitive? Observations on the inertia of the mind. On the improvability of mental capacity.

Chemical questions.

If sulphur and red lead, mixed, are blown out from a bag on to the balls of two jars charged with electricity, but in different states, the positive jar will attract the sulphur, but repel the red lead, whilst the negative jar will attract the red lead and repel the sulphur, and the balls will become coated, this with red lead and that with sulphur. These attractions appear contrary to the laws observed in the voltaic circuit. Query, why? What is the red substance formed when a candle is lighted by a sulphur match? Query, the nature of sounds produced by flame in tubes

Is it possible to imitate, by common electrical induction, the perpetual motion of De Luc's column, as by conveying away two wires, guarded all the way in a tube, to some distance from the ends of a conductor in a state of induction, and attaching leaves to the extremities? or would the induction act along the wires as well as on the conductor, and prevent any effect of attraction? The experiment to be varied, and extended, and applied to explain the column.

What is the nature of the acid formed by the ether lamp and wire? The application of the hydrometer to taking specific gravities of solid bodies.

Do the pith balls diverge by the disturbance of electricity in consequence of mutual induction or not?

When phosphorus is placed in nitrate of silver, is ammonia formed? What is the action of phosphorus on nitrate of ammonia? Query, the nature of the body Phillips burns in his spirit-lamp? He says it is obtained during the distillation of wood. What are the nature and uses of the solution of the tanno-gelatine formed by ammonia?

Effect of the light of peroxide of zinc, heated by a blow-pipe, on a mixture of chlorine and hydrogen.

Distil oxalate of ammonia. Query, results? The nature of the calcareous salt of rhubarb.

General practical observations. Whilst passing through manufactories, and engaged in the observance of the various operations in civilised life, we are constantly hearing observations made by those who find employment in those places, and are accustomed to a minute observation of what passes before them, which are new or frequently discordant with received opinions. These are generally the result of facts, and though some are founded in error, some on prejudice, yet many are true and of high importance to the practical man. Such as come in my way I shall set down here, without waiting for the principle on which they depend; and though three-fourths of them ultimately prove to be erroneous, yet if but one new fact is gathered in a multitude, it will be sufficient to justify this mode of occupying time.

Act against transmutation repealed by 1 William & Mary, st. i. c. 30. Anno quinto Henrici IV. cap. iv.: 'It shall be felony to use the craft of multiplication of gold and silver.'

III.

The original work of this year is seen in the papers he published in the fourth and fifth volumes of the 'Quarterly Journal of Science.' These were on the combustion of the diamond, on the solution of silver in ammonia, on the sulphuret of phosphorus, on some combinations of ammonia with chlorides, and on the sounds produced by flame in tubes.

IV.

Very few of the letters which he wrote in 1818 have been preserved. Directly and indirectly they show how much his time was now occupied. His correspondence with Abbott, with the exception of one letter, consists only of a few lines; when it began it filled almost as many pages. There was a great change in his leisure. There was no change in his friendliness.

The way in which Sir H. Davy ends one of his notes to Faraday at this time shows the position which Faraday had gained. Davy generally wrote to Faraday a very few lines only, upon some business, or about some commission which he wanted him to execute.

FARADAY TO ABBOTT ON THE DEATH OF HIS MOTHER.

'Royal Institution: Friday, February 27, 1818.

'Dear B——, I was extremely shocked at the note I received from you the other day, for the circumstance came on me entirely unexpected. I thought you had all been pretty well, and was regretting that, though I had been negligent in our intercourse, I neither saw nor heard from *you*.

'It is not necessary, B., that I should offer you consolation at this time; you have all the resources necessary within yourself: but I scruple not, though at the risk of reagitating your fatigued feelings, to express my sorrow at the event and condolence with you. It has been my lot of late to see death—not near me, it is true —but around me on all sides, and I have thought and reasoned on it until it has become in appearance harmless, and a very commonplace event. I fear at times that I am becoming too torpid and insensible to the awe that generally, and perhaps properly, accompanies it, but I cannot help it; and when I consider my own weak constitution, the time I have passed, and the probable near approach of that end to all earthly things, I still do not feel that inquietude and alarm which might be expected. It seems but being in another country.

'But I must shorten these reflections. I have not time, nor you, I imagine, serenity to bear them—but philosophise, or, if you please, moralise—the world may laugh as long as it pleases at the *cant* of these terms—so long as they alter not the *things*, they are welcome to their enjoyment. You will find your best resources in reason, and I am sure that, conscious of *that truth*, you have gone to it in distress.

'You say you left many messages at Mr. G.'s: I have not heard them; but I have been little there or anywhere except on business, so that they missed me. I have been more than enough employed. We have been obliged even to put aside lectures at the Institution; and now I am so tired with a long attendance at Guildhall yesterday and to-day, being subpœnaed, with Sir H. Davy, Mr. Brande, Phillips, Aikin, and others, to give chemical information on a trial (which, however, did not come off), that I scarcely know what I say.

'I fear, dear B., that the desultory character of my letter will hurt rather than console your feelings, but I could not refrain longer from acknowledging yours and the pain it gave me.

'Make my kindest remembrances to all; and believe me, dear B., yours as ever,

'M. FARADAY.'

FARADAY TO PROFESSOR G. DE LA RIVE.

'Royal Institution: October 6, 1818.

'Dear Sir,—Your kindness when here in requesting me to accept the honour of a communication with you on the topics which occur in the general progress of science, was such as almost to induce me to overstep the modesty due to my humble situation in the philosophical world, and to accept of the offer you made me. But I do not think I should have been emboldened thus to address you, had not Mr. Newman since then informed me that you had again expressed a wish to him that I should do so; and fearful that you should misconceive my silence, I put pen to paper, willing rather to run the risk of being thought too bold, than of incurring the charge of neglect towards one who had been so kind to me in his expressions.

'My slight attempts to add to the general stock of chemical knowledge have been received with favourable expressions by those around me; but I have, on reflection, perceived that this arose from kindness on their part, and the wish to incite me on to better things. I have always, therefore, been fearful of advancing on what had been said, lest I should assume more than was intended; and I hope that a feeling of this kind will explain to you the length of time which has elapsed between the time when you required me to write and the present moment when I obey you.

'I am not entitled, by any peculiar means of obtaining a knowledge of what is doing at the moment in science, to your attention; and I have no claims in myself to it. I judge it probable that the news of the philosophical world will reach you much sooner through other more authentic and more dignified sources; and my only excuse even for this letter is obedience to your wishes, and not on account of anything interesting for its novelty.

'That my letter may not, however, be entirely devoid of interest, I will take the liberty of mentioning and commending to you a new process for the preparation of gas for illumination.' (He then goes on to describe the preparation of gas from fish oil.)

'I am afraid that with all my reasons I have not been able to justify this letter. If my fears are true, I regret at least it was your kindness drew it from me, and to your kindness I must look for an excuse.

'I am, dear Sir, with great respect, your obedient humble servant, 'M. FARADAY.

The end of one letter from Davy to Faraday is of sufficient interest to be given here.

Rome: October, 1818.

. . . . . .

'Mr. Hatchett's letter contained praises of you which were very gratifying to me; for, believe me, there is no one more interested in your success and welfare than your sincere well-wisher and friend,

'H. DAVY.'

In 1819 the medium through which Faraday's nature and the progress of his knowledge are seen in one respect differs from that of the previous year. During an excursion into Wales in July he kept a journal. In it he shows his love of beautiful scenery, his power of describing it, his close observation of the metal and slate works of the country, and the overflowing kindness of his nature. The notes of one lecture which he gave at the City Philosophical Society have been kept. The subject was the forms of matter. It shows the freedom and breadth of his thoughts, his desire 'to avoid the whisperings of fancy,' his belief that facts are the only things which we are *sure* are worthy of trust. In one part of his lecture he brings forward that fourth or radiant state of matter which has now passed out of thought. And he states that at this period he inclined to the opinion of the immaterial nature of heat, light, electricity, &c. He ended thus: 'All I wish to point out is the necessity to preserve the mind from philosophical prejudices. The man who wishes to advance in knowledge should never of himself fix obstacles in the way.'

His note-book contains nothing of interest. His published papers were many, but none of great importance. There is only one letter to Abbott, and that shows how much his time was occupied. It appears from the letters of Sir H. Davy, that the Herculaneum manuscripts might have been subjected to Faraday's manipulation at Naples if he could have left the Royal Institution.

I.

JOURNAL OF A WALKING TOUR IN WALES.

THE COPPER WORKS OF SWANSEA, THE MINES OF ANGLESEA, AND SLATE MINES OF BANGOR.

The Regulator is an excellent coach. I mounted the top of it at the White Horse, Piccadilly, about a quarter past five on Saturday morning, July 10, and it set me down in Bristol about ten o'clock the same evening. On getting up next morning, Sunday, July 11, I troubled the waiter for an early breakfast, and then, with an old ostler for my guide, set off to see Clifton and Hot Wells.

Having returned to the inn and bade my guide good morning, I found I had still time to spare for a walk; the mail not leaving town until past twelve o'clock. On the quay I found a very respectable man preaching in a very respectable manner from the top of a doghouse to a number of persons; among whom were many seamen and poor women extremely neat and clean in their dress. After general exhortations, he addressed himself to the seamen more particularly, and called their attention in very good and strong language to their mode of life, their dangers, their resources, and their general moral habits. He did not hesitate to censure them very strongly; but yet he succeeded in commanding so much respect as to give no offence. The men around him, mates and common seamen, were extremely attentive; many were uncovered, though the preacher himself was not. And in listening, it did not seem the mere effect of habit or duty, but a strong and interested attention. They chid some unruly boys in a way which showed their seriousness and decorum, and I left them an earnest preacher and an attentive audience.

After an early breakfast, Monday morning, July 12, at Cardiff, I took a postchaise and proceeded on to Merthyr. In the afternoon I rambled with Mr. Guest's agent over the works at Dowlais. I was much amused by observing the effect the immensity of the works had on me. The operations were all simple enough, but from their extensive nature, the noise which accompanied them, the heat, the vibration, the hum of men, the hiss of engines, the clatter of shears, the fall of masses, I was so puzzled I could not comprehend them except very imperfectly. The mind was drawn to observe effects rather for their novelty than for their importance, and it was only when by going round two or three times I could neglect to listen to sounds at first strange, or to look at rapid motions, that I could readily trace the process through its essential parts, and compare easily and quickly one part with another.

Saturday, 17*th*.—We crossed Neath bridge about nine o'clock. An innkeeper of the town, who was entering it on horseback, saluted us in passing, and put his card into our hands; and so slight was the impulse which directed our course, that this was sufficient to take us to his house. I think there cannot be a more pleasurable feeling belonging to existence, than that which a man has when no artificial circumstance confines him. Every place is a home, every being a friend. He is always in possession of his object, he runs no risk from disappointment. Wherever we went it was the same to us; wherever we went we were sure of novelty and pleasure.

Monday, 19*th*.—Proceeding onwards into Brecknockshire, we suddenly heard the roar of water where we least expected it, and came on the edge of a deep and woody dell. Entering among the trees, we scrambled onwards after our guide, tumbling and slipping and jumping and swinging down the steep sides of the dingle, sometimes in the path of a running torrent, sometimes in the projecting fragments of slate, and sometimes where no path or way at all was visible. The thorns opposed our passage, the boughs dashed their drops in our faces, and stones frequently slipped from beneath our feet into the chasm below, in places where the view fell uninterrupted by the perpendicular sides of the precipices. By the time we had reached the bottom of the dingle, our boots were completely soaked, and so slippery that no reliance could be placed on steps taken in them. We managed, however, very well, and were amply rewarded by the beauty of the fall which now came in view. Before us was a chasm inclosed by high perpendicular and water-worn rocks of slate, from the sides of which sprang a luxuriant vegetation of trees, bushes, and plants. In its bosom was a basin of water, into which fell from above a stream divided into minute drops from the resistance of its deep fall. Here and there lay trunks of trees which had been brought down by the torrent—striking marks of its power—and the rugged bed of shingles and rocky masses further heightened the idea other objects were calculated to give of the force it possessed when swelled by rains. We stepped across the river on a few tottering and slippery stones placed in its bed, and passing beneath the overhanging masses ran round on projecting points, until between the sheet of water and the rock over which it descended; and there we remained some time, admiring the scene. Before us was the path of the torrent, after the fine leap which it made in this place; but the abundance of wood hid it ere it had proceeded many yards from the place where it fell. No path was discernible from hence, and we seemed to be inclosed on a spot from whence there was no exit, and where no cry for help could be heard because of the torrent-roar.

The effect of the wind caused by the descent of the stream was very beautiful. The air carried down by the stream, the more forcibly in consequence of the minute division of the water, being resisted by the surface of the lake beneath, passed off in all directions from the fall, sweeping many of the descending drops with it. Between us and the fall the drops fell brilliant and steady till within a few inches of the bottom, when receiving a new impulse they flew along horizontally, light and airy as snow. A mist of minute particles arose from the conflicting waters, and being driven against the rocks by the wind clothed them with moisture, and created myriads of miniature cascades, which falling on the fragments beneath polished them to a state of extreme slipperiness.

The fall is called Scwd-yv-hên-rhyd, or Glentarec, and is produced by the descent of the river Hên-rhyd. It is called 300 feet high, but is really only 105. The river afterwards proceeds into the Towey Vale, and discharges itself at Swansea into the Channel.

Tuesday, 20*th*.—After dinner I set off on a ramble to Melincourt, a waterfall on the north side of the valley, and about six miles from our inn. Here I got a little damsel for my guide who could not speak a word of English. We, however, talked together all the way to the fall, though neither knew what the other said. I was delighted with her burst of pleasure as, on turning a corner, she first showed me the waterfall. Whilst I was admiring the scene, my little Welsh damsel was busy running about, even under the stream, gathering strawberries. On returning from the fall I gave her a shilling that I might enjoy her pleasure: she curtsied, and I perceived her delight. She again ran before me back to the village, but wished to step aside every now and then to pull strawberries. Every bramble she carefully moved out of the way, and ventured her bare feet to try stony paths, that she might find the safest for mine. I observed her as she ran before me, when she met a village companion, open her hand to show her prize, but without any stoppage, word, or other motion. When we returned to the village I bade her good-night, and she bade me farewell, both by her actions and I have no doubt her language too.

Sterne may rise above Peter Pastoral and Stoics above Sterne, in the refined progress of human feeling and human reason, but he who feels and enjoys the impulses of nature, however generated, is a man of nature's own forming, and has all the dignity and perfection of his race, though he may not have adopted the refinements of art. I never felt more honourable in my own eyes than I did this evening, whilst enjoying the display this artless girl made of her feelings.

The evening was beautiful; a short fine sunset ornamented the heavens with a thousand varying tints, and my walk home was delightful.

Saturday, 24th.—We departed from the Devil's Bridge, the waiter having assured us we should easily find our way over the mountains to Machynleth. Now this sounded very smooth and fair, but no account was taken of the following circumstances: 1, no roads; 2, no houses; 3, no people; 4, rivers but no bridges; and 5, plenty of mountains.

When we were within five miles of Machynleth the scenery began to change. On our left we had a fine view of Cardigan Bay, beneath us was the mouth of the river Dovey or Dyfi opening into it, and in the distance the extreme northern points of the bay in Carnarvonshire. Before us was Cader Idris rising above a host of mountains assembled at his base, and separated from us by a broad, deep, well-cultivated and wooded valley. On our right were mountains near at hand, part of the same chain as those on which we stood, and between their summits every now and then Plynlimmon appeared. We descended from our exalted station along a rugged path into the vale beneath, and soon entered amongst wood on the sides of the dells. The scenery became more and more enchanting as we proceeded, equalling all the cultivated beauties of Hafod, and surpassing them in the introduction of peasants' huts of the finest form and state for a picture.

I wanted a little alcohol, and having found out a doctor's shop and a spruce doctor's man, got some. I then asked for a little spirits of salts, hoping I could have it in a glass-stoppered bottle. The man found me a bottle, having emptied one of his preparations out of it, and would then have poured in acid; but it was not the acid I wanted, and I again mentioned spirits of salts to him, willing to allow everything to the possibility of his ignorance of the scientific name, but at the same time adding muriatic acid, to save his credit if possible. He now seemed to understand me, and reaching down another bottle again prepared to pour, but I stopped him. 'It is muriatic acid that I want.' 'This is muriatic acid, sir.' 'No, that is nitric acid.' 'They are the same, sir.' 'Oh no, there is a little difference between them, and one will not do for me so well as the other.' I then endeavoured to explain that the one came from nitre, the other from common table salt. He comprehended a difference between these two bodies, but not between their acids; and he brought out a pharmacopœia, and opening it at muriatic acid, uttered the Latin name and synonyms fluently and with great emphasis, endeavouring thus to prove to me the two were alike. I was really ashamed to correct the doctor, and if I had not been under the necessity of vindicating my contradiction of him, should have left him in ignorance. However, at last I made him comprehend from his own book that there was something like a difference between these acids, but I don't think he shut the book much improved by the affair. I could scarcely afterwards look at the man. If he had any feeling—

and he appeared to have a considerable stock of pride—he must have felt himself extremely lowered in the eyes of strangers, and before his own companion who was standing by. I began to rummage his bottles for muriatic acid myself, but I must do him the justice to say that he first found out what little they had (about an ounce), and that he really compared it with the nitric acid—I hope for information, though his object professedly was to show me how like they were.

Is it not strange that a man so ignorant of his profession should still appear respectable in it, or that one so incompetent should be entrusted with the health and lives of his fellow-creatures? Had I seen nothing more than his haughty dictatorial behaviour to a poor woman who came in with a prescription and a bottle in her hand, I should have concluded him to be a man who had attained the utmost knowledge of and confidence in his art; seeing what I did, I cannot enough condemn the being who with such ignorance still apes the importance of highest wisdom, and who, without a knowledge even of the first requisites to an honourable but dangerous profession, assumed to himself its credit and its power, and dashed at once upon human life with all the means of destruction about him, and the most perfect ignorance of their force.

Sunday, July 26th.—Ascent of Cader Idris.

The thunder had gradually become more and more powerful, and now rain descended. The storm had commenced at the western extremity of the valley, and rising up Cader Idris traversed it in its length, and then passing over rapidly to the south-east, deluged the hills with rain. The waters descended in torrents from the very tops of the highest hills in places where they had never yet been observed, and a river which ran behind the house into the lake below rose momentarily, overflowed its banks, and extended many yards over the meadows. The storm then took another direction, passing over our heads to the spot in the west at which it had commenced, and having been very violent in its course, seemed there to be exhausted and to die away. The scene altogether was a very magnificent one—the lightning's vivid flash illuminated those parts which had been darkened by its humid habitation, and the thunder's roar seemed the agonies of the expiring clouds as they dissolved into rain; whilst the mountains in echoes mocked the sounds, and laughed at the fruitless efforts of the elements against them.

The wind rose in all directions, and I observed here, as has often been observed elsewhere, that the track of the storm seemed quite independent of it: the storm seeming rather to direct the wind than the wind the storm.

I rambled out as soon as the rain and hail would permit me, but it is impossible to describe the various characters and beauties of the fine views presented by this beautiful vale during and after the storm. Every fresh cloud, every change in the atmosphere, varied the combination of form and character belonging to the scene, and presented, as it were, a new one; and I enjoyed all the pleasures of extensive and varying scenery without moving from my situation.

Wet-footed and fatigued, we were obliged to stop when about one-third of the distance from the summit, though on a very precarious tenure; but now, more to embarrass us, clouds began to roll towards the summit, and rain descended. We again hastened our exertions in order to get to the top and see where we were before all became hid in obscurity, and we at last gained the corner in a state of great exhaustion. The most dangerous part of the road as regarded our feet was now passed, but, enveloped in cloud, every part was dangerous. Nevertheless we pressed on. The mountain rose before us to a great height, terminating above in points, and the space between us and them was covered with large loose boulders. We stepped over them as quickly as possible, both for the sake of seeing from the top, before the storm came on, in what direction we were, and of getting shelter; but the weather was too hard for us—in a few moments we could see nothing ten yards off, and only knew we were going up from the inclined surface.

In a short time we got to a more level space, and hastening on to its edge endeavoured to look before us; but the cloud and rain hid all, and all we could learn was that a tremendous chasm was there, indicated by the whiteness of the cloud that way, for on looking down we seemed to look into a sky. We retreated to a cleft in the rock for shelter from the wind and rain, meaning to wait some time in hopes the clouds would blow over. I was terribly alarmed here by supposing I had lost the compass: however, the compass was found at last, and then we got over one difficulty.

Advancing merely for the sake of preserving ourselves in motion, we came to tracks among the fragments, which, as they led to higher ground than that we stood upon, we followed, and after winding some time got to the highest cliff, and there found the hut made by the guides for shelter from storms. We did not stop there long, and had not left many minutes before the clouds began to disperse. We ran to the edge of the precipice to see where we were, and obtained a glimpse of a very magnificent scene; but it was too transient, and we still had to wait for information. In a short time, however, it became clear, and we could perceive that the mountain expanded above into a ridge in some places a hundred yards wide, here and there covered with fragments of the rock, but in general green from grass. We could not as yet see into the valley, though the top was pretty free from clouds. The vapours remained beneath a long time, and it was pleasing to observe how the wind swept over the edge of the precipice, levelling the mist above its height, but leaving the valleys full. We got a peep over all into Cardigan Bay, and at the mountains in the distance. It was very sublime, and the mixture of air and earth thus presented was equal, I think, in effect to anything that can be imagined. After proceeding about three miles, we began to be assured we were on our way to Dolgelly.

Friday, 30*th*.—Breakfasted this morning at the Mitre Tavern, next door to the cathedral of this mighty city of Bangor. I am no judge of civic propriety, but I could not help taking Bangor for the caricature of a city, when told that it claimed the right of being called one. As for its cathedral, I was afraid to look at that, lest, from the glimpse I had already caught of it, I should take it for a caricature too, and I wished to give reverence for the name's sake.

We started at a good pace to see the slate quarries, about six miles from Bangor.

We now began to see the quarries at intervals from amongst the trees, like a number of hills of rubbish, on the side of a mountain before us, and their appearance increased our eagerness to be at them.

We had to make our way round and between several high hills of refuse slate before we got fairly into the works; but when there, we were charmed with the novel and strange appearances of things. The splintery character of all about us, the sharp rocky projections above us, the peculiar but here general colour of the rock, together made up an appearance unlike anything I had seen before. We pushed on, boldly passing men and offices, and went up inclined ways and along railroads towards the explosions we heard a little way off. After having seen two or three very curious places, we tempted a man to leave his work and show us the road to the most interesting parts of the quarry, and he took us among the cliffs, where we almost repented we had asked to go. Smooth perpendicular planes of slate, many many feet in height, depth, and width, appeared above and below; in all directions chasms yawned, precipices frowned; and the path which conducted amongst and through these strange places was sometimes on the edge of a slate splinter not many inches wide, though raised from the cliffs beneath into mid air. We then mounted, and at last gained a kind of slate promontory, which had been left projecting across the quarry. It was narrow, but walled on both edges. From hence we had a kind of bird's-eye view of the excavations and workings, and saw the men like pigmies below pursuing their various objects. It was certainly a very singular scene, and is like nothing else that I know of. Natural precipices do not convey the idea excited here, because they are in part rounded by the weather, and their smaller parts are generally somewhat nodular or blunt, and, besides, they are modified in colour by the soil that lodges on and the vegetation that covers them. But here every fracture, whether large or small, presented sharp angles ; the fine sober colour was of the utmost freshness, and, in opposition to usual arrangement, the sides were the smooth and flat places, the bottoms being the rough irregular part, for the strata here are nearly perpendicular. All over the place were scattered men, sometimes sitting across a little projection starting from the sharp edge, or clinging on to a half-loosened splinter of the slate, and employed in making holes, tamping and blasting the rock. Railways wound in every direction into the works, and waggons were continually moving about in the lively scene. Just before us they were going to blast, and they motioned us away from the place to be out of danger. The explosion did not, however, scatter the fragments, but it made a noble roar.

We took the coach to Llangollen. The shades of evening now began to gather over us, and we all sank as if by agreement into a very quiet sober state, in which we should have continued perhaps the rest of our ride had not the expertness and agility of a blind woman roused us. Poor Bess was waiting in the road for the coaches—for there were two, ours and another immediately behind it—and on hearing the noise of the wheels, stepped a little on one side. When they were up to her she ran to the hinder part, and feeling for the wheels or the steps, got hold of the irons, and immediately mounted, spite of the coach's motion or her blindness. She and the guard appeared to be old friends, but we soon found her object was to sell her goods. She pulled out some Welsh wigs knitted by herself, and offered them to the passengers behind for sale, pleading her blindness as a strong reason why they should purchase. I bought one of the Welsh wigs in remembrance of the old woman, and gave her one shilling and sixpence for it. She then pulled out some socks; on which I scolded her for not producing them before. They were a shilling a pair, and I wanted to take them instead of the wig; but she was unwilling, even though I offered the eighteen-penny wig for the shilling socks; and we found afterwards from the guard, to whom she confessed her reason, that the wigs were her own knitting, but the socks she only sold for a neighbour. She then stretched across the coach to the front passengers, but they would not buy; and afterwards she clung to the side of the vehicle, standing on the spring between the wheel and the body of the coach, and peeping (query) in at the windows, offered her articles for sale there: but they were boys only, and they did not want socks or Welsh wigs.

Finding further stay on our coach useless, she descended, and avoiding the horses of the coach immediately behind us, got round it and mounted in the same manner as she had done with us. But I am afraid all her exertions did not gain her any further success at this time, for she soon after got down and went off home.

Saturday, 31*st*.—Llangollen. We had time this morning to enjoy the inn we had entered, and which possesses a very high character for cleanliness, attention, and comfort. We certainly found it so, and entirely free from the inconveniences which inns have in general either more or less. Whilst at breakfast, the river Dee flowing before our windows, the second harper I have heard in Wales struck his instrument, and played some airs in very excellent style. I enjoyed them for a long time, and then, wishing to gratify myself with a sight of the interesting *bard*, went to the door and beheld—the *boots!* He, on seeing me open the door, imagined I wanted something, and, quitting his instrument, took up his third character of *waiter.* I must confess I was sadly disappointed and extremely baulked. Even at Bethgellert they had a good-looking blind old man, though he played badly; and now, when I heard delightful sounds, and had assured myself the harper was in accordance with the effect he produced, he sank on a sudden, many many stages down, into a common waiter. Well—after all, I certainly left Llangollen regretting the harp less because of the person who played it.

This year a lecture, On the Forms of Matter, was given at the City Philosophical Society: in it he shows his views regarding matter and force at this time.

'In the constant investigation of nature pursued by curious and inquisitive man, some causes which retard his progress in no mean degree arise from the habits incurred by his exertions; and it not unfrequently happens, that the man who is the most successful in his pursuit of one branch of philosophy thereby raises up difficulties to his advancement in another.

'Necessitated as we are, in our search after the laws impressed upon nature, to look for them in the effects which are their aim and end, and to read them in the abstracted and insulated phenomena which they govern, we gradually become accustomed to distinguish things with almost preternatural facility; and induced by the ease which is found to be afforded to the memory and other faculties of the mind, division and subdivision, classification and arrangement, are eagerly adopted and strenuously retained.

'Much as the present stage of knowledge owes to this tendency of the human mind to methodise, and therefore to facilitate its labours, still it may complain that in some directions it has been opposed and held down to error by it. All method is artificial and all arrangement arbitrary. The distinction we make between classes, both of thoughts and things, are distinctions of our own; and though we mean to found them on nature, we are never certain we have actually done so. That which appears to us a very marked distinctive character may be really of very subordinate importance, and where we can perceive nothing but analogies and resemblances, may be concealed nature's greatest distinctions.

'The evil of method in philosophical pursuits is indeed only apparent, and has no real existence but in the abuse. But the system-maker is unwilling to believe that his explanations are not perfect, the theorist to allow that incertitude hovers about him. Each condemns what does not agree with *his method*, and consequently each departs from nature. And unfortunately, though no one can conceive why another should presume to bound the universe and its laws by his wild and fantastic imaginations, yet each has a reason for retaining and cherishing his own.

'The disagreeable and uneasy sensation produced by incertitude will always induce a man to sacrifice a slight degree of probability to the pleasure and ease of resting on a decided opinion; and where the evidence of a thing is not quite perfect, the deficiency will be easily supplied by desire and imagination. The efforts a man makes to obtain a knowledge of nature's secrets merit, he thinks, their object for their reward; and though he may, and in many cases must, fail of obtaining his desire, he seldom thinks himself unsuccessful, but substitutes the whisperings of his own fancy for the revelations of the goddess.

'Thus the love a man has for his own opinion, his readiness to form it on uncertain grounds rather than remain in doubt, and the necessity he is under of referring to particular and individual examples in illustration of his views of nature, all tend to the production of habits of mind which are partial and warped. These habits it is which give rise to the difference of opinion in men on every possible subject. All parts of the system both of the moral and natural world are constant in their natures, presenting the same appearances at all times and to all men. But we cannot perceive them in all their bearings and relations; we view them in different states and tempers of mind, and we hasten to decide upon them. Hence it happens, that a judgment is made for future use which not only differs in different individuals, but, unfortunately, from truth itself.

'As it regards natural philosophy, these bad, but more or less inevitable, effects are perhaps best opposed by cautious but frequent generalisations. It is true that with the candid man experience will do very much, and after having found in some instances the necessity of altering previous opinions, he will retain a degree of scepticism in future on all those points which are not proved to him. But generalisations will aid the efforts a man makes to free himself from erroneous ideas and prejudices; for, presenting to us the immense family of facts ranged according to the relationships of the individuals, it makes evident many analogies and distinctions which escape the mind when engaged on each separately, and corrects those errors consequent on partial views.

'We are obliged, from the confined nature of our powers, to consider of but one thing at a time. Generalisation compensates in part the resulting inconveniences, and in an imperfect way places many things before us; and the more carefully this is done the more accurately our partial notions are corrected.

'Ultimately, however, facts are the only things which we are *sure* are worthy of trust. All our theories and explanations of the laws which govern them, whether particular or general, are necessarily deduced from insufficient data. They are probably most correct when they agree with the greatest number of phenomena, and when they do not appear incompatible with each other. The test of an opinion is its agreement in association with others, and we associate most when we generalise.

'Hence I should recommend the practice of generalising as a sort of parsing in philosophy. It occasions a review of single opinions, requires a distinct impression of each, and ascertains their connection and government. And it is on this idea of the important use that may be made of generalisation, that I venture to propose for this evening a lecture on the general states of matter.

'Matter defined—essential and secondary properties.

'Matter classed into four states—solid, liquid, gaseous,

and radiant—which depend upon differences in the essential properties.

. . . . . .

'Radiant state.—Purely hypothetical. Distinctions.

'Reasons for belief in its existence. Experimental evidence. Kinds of radiant matter admitted.

'Such are the four states of matter most generally admitted. They do not belong to particular and separate sets of bodies, but are taken by most kinds of matter; and it will now be found necessary, to a clear comprehension of their nature, to notice the phenomena which cause and accompany their transition into each other.

'Some curious points arise respecting the changes in the forms of matter, which, though not immediately applicable to any convenient or important use, claim our respect as buddings of science which at some future period will be productive of much good to man. Of the bodies already taken and presented in various forms in illustration of this part of our subject, some have evinced their Protean nature in the production of striking effects; others there are which, being more constant to the states they take on, suffer a conversion of form with greater difficulty; and others, again, have as yet resisted the attempts made to change their state by the application of the usual agencies of heat and cold. By the power of heat all solid bodies have been fused into fluids, and there are very few the conversion of which into a gaseous form is at all doubtful. In inverting the method, attempts have not been so successful. Many gases refuse to resign their form, and some fluids have not been frozen. If, however, we adopt means which depend on the rearrangement of particles, then these refractory instances disappear, and by combining substances together we can make them take the solid, fluid, or gaseous form at pleasure.

'In these observations on the changes of state, I have purposely avoided mentioning the radiant state of matter, because, being purely hypothetical, it would not have been just to the demonstrated parts of the science to weaken the force of their laws by connecting them with what is undecided. I may now, however, notice a curious progression in physical properties accompanying changes of form, and which is perhaps sufficient to induce, in the inventive and sanguine philosopher, a considerable degree of belief in the association of the radiant form with the others in the set of changes I have mentioned.

'As we ascend from the solid to the fluid and gaseous states, physical properties diminish in number and variety, each state losing some of those which belonged to the preceding state. When solids are converted into fluids, all the varieties of hardness and softness are necessarily lost. Crystalline and other shapes are destroyed. Opacity and colour frequently give way to a colourless transparency, and a general mobility of particles is conferred.

'Passing onward to the gaseous state, still more of the evident characters of bodies are annihilated. The immense differences in their weights almost disappear; the remains of difference in colour that were left are lost. Transparency becomes universal, and they are all elastic. They now form but one set of substances, and the varieties of density, hardness, opacity, colour, elasticity and form, which render the number of solids and fluids almost infinite, are now supplied by a few slight variations in weight, and some unimportant shades of colour.

'To those, therefore, who admit the radiant form of matter, no difficulty exists in the simplicity of the properties it possesses, but rather an argument in their favour. These persons show you a gradual resignation of properties in the matter we can appreciate as the matter ascends in the scale of forms, and they would be surprised if that effect were to cease at the gaseous state. They point out the greater exertions which nature makes at each step of the change, and think that, consistently, it ought to be greatest in the passage from the gaseous to the radiant form; and thus a partial reconciliation is established to the belief that all the variety of this fair globe may be converted into three kinds of radiant matter.

'There are so many theoretical points connected with the states of matter that I might involve you in the discussions of philosophers through many lectures without doing justice to them. In the search after the *cause* of the changes of state of bodies, some have found it in one place, some in another; and nothing can be more opposite than the conclusions they come to. The old philosophers, and with them many of the highest of the modern, thought it to be occasioned by a change either in the motion of the particles or in their attractive power; whilst others account for it by the introduction of another kind of matter, called heat, or caloric, which dissolves all that we see changed. The one set assume a change in the *state* of the matter already existing, the other create a new kind for the same end.

'The nature of heat, electricity, &c., are unsettled points relating to the same subject. Some boldly assert them to be matter; others, more cautious, and not willing to admit the existence of matter without that evidence of the senses which applies to it, rank them as qualities. It is almost necessary that, in a lecture on matter and its states, I should give you my own opinion on this point, and it inclines to the immaterial nature of these agencies. One thing, however, is fortunate, which is, that whatever our opinions, they do not alter nor derange the laws of nature. We may think of heat as a property, or as matter: it will still be of the utmost benefit and importance to us. We may differ with respect to the way in which it acts: it will still act effectually, and for our good; and, after all, our differences are merely squabbles about words, since nature, our object, is one and the same.

'Nothing is more difficult and requires more care than philosophical deduction, nor is there anything more adverse to its accuracy than fixidity of opinion. The man who is certain he is right is almost sure to be wrong, and he has the additional misfortune of inevitably remaining so. All our theories are fixed upon uncertain data, and all of them want alteration and support. Ever since the world began, opinion has changed with the progress of things; and it is something more than absurd to suppose that we have a sure claim to perfection, or that we are in possession of the highest stretch of intellect which has or can result from human thought. Why our successors should not displace us in our opinions, as well as in our persons, it is difficult to say; it ever has been so, and from analogy would be supposed to continue so; and yet, with all this practical evidence of the fallibility of our opinions, all, and none more than philosophers, are ready to assert the real truth of their opinions.

'The history of the opinions on the general nature of matter would afford remarkable illustrations in support of what I have said, but it does not belong to my subject to *extend upon it.* All I wish to point out is, by a reference to light, heat, electricity, &c., and the opinions formed on them, the necessity of cautious and slow decision on philosophical points, the care with which evidence ought to be admitted, and the continual guard against philosophical prejudices which should be preserved in the mind. The man who wishes to advance in knowledge should never of himself fix obstacles in the way.'

III.

The few notes in his commonplace-book for this and the following year scarcely relate to science at all.

Spoiled mutton—a parody on 'The Rose had been washed.'

The figures 1 2 3, a fable.

Iron columns in arcade round the Opera House cracked and broke in two.

Leather fluid. Agricultural chemistry, Lecture 2nd, clearing land of stones.

The Swiss song, 'Ranz des Vaches,' imitated.

Pastilles. Preaching Trade. Remedy for warts, juice from inner surface of broad-bean shells. Water ices and ice cream—alliteration, monosyllabic writing.

IV.

He had nineteen notices and papers in the 'Quarterly Journal of Science,' volumes six and seven: the most important were on sirium or vestium, on the action of boracic acid on turmeric, on gallic acid, tannin, &c.;

on the separation of manganese from iron, on carburetted hydrogen, on nitrous oxide, on analysis of wootz or Indian steel, and on experimental observations on the passage of gases through tubes. In this last paper he continued the observations he had made in the previous year. His chief conclusion was that there was much more work to do on this subject, a result which the discoveries of Professor Graham in 1834 and 1849 splendidly confirmed.

FARADAY TO ABBOTT.

'Royal Institution: April 27, 1819.

. . . . . .

'You will be aware that the business of the Institution must press hard upon me at this time and during the whole of the lectures. To this is added much private employment, which will not admit of neglect. With reference to my evenings, they are thus arranged:—On Monday evening there is a scientific meeting of members here, and every other Monday a dinner, to both of which my company is requisite. On Tuesday evening I have a pupil, who comes at six o'clock and stops till nine, engaged in private lessons. On Wednesday the Society requires my aid. Thursday is my only evening for accidental engagements. Friday, my pupil returns and stops his three hours; and on Saturday I have to arrange my little private business. Now you will see that, except on Tuesday and Friday after 9 o'clock, I have no evening but Thursday for anything that may turn up. Now for Thursday night I have engaged with a party to see Matthews, whom as yet I have not seen; but on the Thursday following I shall expect you if you can find it convenient.

'I must now hasten to my crucibles, so for the present adieu.

'Yours very sincerely,
'M. FARADAY.'

Early in February, 1819, Sir H. Davy wrote to Faraday from Rome,—'I have sent a report on the state of the MSS. to our Government, with a plan for the undertaking of unrolling; one part of the plan is to employ a chemist for the purpose at Naples: should they consent, I hope I shall have to make a proposition to you on the subject.'

In May, Sir H. Davy writes from Florence:—

'It gives me great pleasure to hear that you are comfortable at the Royal Institution, and I trust that you will not only do something good and honourable for yourself, but likewise for science.

'I am, dear Mr. Faraday, always your sincere friend and well-wisher,

'H. DAVY.'

At the end of the year Sir H. Davy again writes on this subject:—

'Could you have left the Royal Institution for a few months or a year, and have been secure of returning to your situation, I should have strongly recommended to you the employment at Naples. This indeed is still open, for the person I have engaged as operator is hired by the month. When I have seen my way a little as to the time the MS. operations will demand, I will write to you.'

CHAPTER V.

HIGHER SCIENTIFIC EDUCATION AT THE INSTITUTION—MARRIAGE —FIRST PAPER IN THE 'PHILOSOPHICAL TRANSACTIONS.'

1820. ÆT. 28–29.

IF it were desirable to fix any date when the scientific education of Faraday might be said to have ended, and his work as an educated man of science might be said to have begun, it would be at the beginning of this period.

For seven years as the private assistant of Davy, and as assistant in the laboratory and lecture-room at the Royal Institution, Faraday had now served his apprenticeship to science. He had begun a most laborious original investigation with Mr. James Stodart on the alloys of steel, which he was now about to publish. He had already had thirty-seven notices and papers printed in the 'Quarterly Journal of Science' (one or two of them were of great importance to science), and he had given his first course of lectures on chemistry at the City Philosophical Society with great success as a speaker and experimenter.

But highly as Faraday was at this time educated, and much as he had done, he was as yet only at the beginning of a still higher education. It was not until the eleventh year from this date that his first paper 'On Experimental Researches in Electricity' was published.

In other words, he took eighteen years to educate himself for the great scientific work which he had to do.

The progress of this education, the reputation which he obtained, the traits of his character, and the course of his life during this period, will be now shown, (1) in the works he published; (2) in the lectures he gave; (3) in the honours paid to him; (4) in the letters which he wrote and received.

The events of this year may in some respects be considered the most important in the life of Faraday. His acquaintance with Mr. Tatum in 1810 in Fleet Street, and his introduction in 1812 to Sir Humphry Davy, had fixed his course in life, and had made him a scientific man. His engagement this year to be married to Sarah, the third daughter of Mr. Barnard of Paternoster Row, an elder of the Sandemanian Church, made him a happy man for forty-seven years.

Before proceeding to trace the progress of his education in science, by his publications and letters, it will be well to show his mind and character by means of a short journal, and the letters which he wrote in this and the following year on the subject of his proposal and his marriage. In this, as in every other act of his life, he laid open all his mind and the whole of his character, and he kept back none of his thoughts; and what can here be made known can scarcely fail to charm every one by its loveliness, its truthfulness, and its earnestness.

On two or three occasions Faraday had made notes in his commonplace-book of passages against love. His friend, Mr. Edward Barnard, saw this book, and spoke of these passages to his sister Sarah. Thus she knew that Faraday's thoughts were not in the way of matrimony.

On the 5th of July, 1820, Faraday wrote to Miss Sarah Barnard :—

'Royal Institution.

'You know me as well or better than I do myself. You know my former prejudices, and my present thoughts—you know my weaknesses, my vanity, my whole mind; you have converted me from one erroneous way, let me hope you will attempt to correct what others are wrong.

. . . . . .

'Again and again I attempt to say what I feel, but I cannot. Let me, however, claim not to be the selfish being that wishes to bend your affections for his own sake only. In whatever way I can best minister to your happiness either by assiduity or by absence, it shall be done. Do not injure me by withdrawing your friendship, or punish me for aiming to be more than a friend by making me less; and if you cannot grant me more, leave me what I possess, but hear me.'

Miss Barnard showed this letter to her father, and he, instead of helping her to decide, said that love made philosophers say many foolish things. Her own youth and fear made her hesitate to accept Faraday, and she left London with her sister, Mrs. Reid, for Ramsgate, in order to postpone any immediate decision.

Faraday thus writes in a journal which he made of the following week :—

'*July 29th*.—I made up my mind yesterday afternoon to run all risks of a kind reception at Ramsgate, and force myself into favourable circumstances if possible.' On the evening of his arrival there he says, 'I was in strange spirits, and had very little command over myself, though I managed to preserve appearances. I expressed strong disappointment at the look of the town and of the cliffs, I criticised all around me with a malicious tone, and, in fact, was just getting into a humour which would have offended the best-natured person, when I perceived that, unwittingly, I had, for the purpose of disguising the hopes which had been raised in me so suddenly, and might have been considered presumptuous, assumed an appearance of general contempt and dislike. The moment I perceived the danger of the path on which I was running, I stopped, and talked of home and friends.'

Two days afterwards he says, 'During a walk the conversation gradually became to me of the most pensive cast, and my mind was filled with melancholy thoughts. We went into a mill and got the miller to show us the machinery; thus seeking mechanical means of changing the subject, which, I fear, weighed heavy on both of us. But still our walk continued to have a very sombre, grave cast with it; and when I sat down in the chair at home, I wished for a moment that memory and sensation would leave me, and that I could pass away into nothing. But then pride came to my help, and I found that I had at least one independent auxiliary left, who promised never to desert me whilst I had existence.'

A day or two afterwards Faraday and Miss Barnard went to Dover. They ascended the shaft, and going to a higher part of the hill, 'we came within view of Shakespeare's Cliff. The scene from hence was very fine, and quite beyond anything I had seen among chalk strata before. The cliffs rose like mountains, not with the exactly perpendicular sides, the flat top, the uncovered surface of the chalk round Thanet, but with steep and overhanging declivities, carrying the peaks to an immense height, with summits and ridges towering in the air, with sides beautifully broken into the rude grandeur and variety of mountain forms clothed with a varied and luxuriant vegetation, and supporting even trees in the dents and crevices with which they abounded. At the foot of these cliffs was the brilliant sparkling ocean, stirred with life by a fresh and refreshing wind, and illuminated by a sun which made the waters themselves seem inflamed. On its surface floated boats, packets, vessels, beating the white waves, and making their way against the feigned opposition of the waters. To our left lay Dover, with its harbour and shipping equally sheltered and threatened by the surrounding hills; and opposite were the white cliffs of the French coast, the dim outlines and thin shades of which just enabled us to guess that they also rose in irregular forms, and were broken into variety of surface.

'The whole was beautiful, or magnificent, as the mind received its tone from successive thoughts, and almost became sacred when the eye wandered towards the arch cliff, for there Shakespeare's spirit might be fancied sitting on the very verge, absorbed in the contemplation of its grandeur. Then imagination would figure the bark, the rock, the buoy, the very lark to whom that mortal has given an existence that will end only with mortality.

. . . . . .

'I can never forget this day. Though I had ventured to plan it, I had had little hope of succeeding. But when the day came, from the first waking moment in it to the last it was full of interest to me: every circumstance bore so strongly on my hopes and fears that I seemed to live with thrice the energy I had ever done before. . . . .

'But now that the day was drawing to a close, my memory recalled the incidents in it, and the happiness I had enjoyed; and then my thoughts saddened and fell, from the fear I should never enjoy such happiness again.

. . . . . .

'I could not master my feelings or prevent them from sinking, and I actually at last shamed myself by moist eyes. . . . It is certainly strange that the sincerity and strength of affection should disable me from judging correctly and confidently of the heart I wish to gain, and adopting the best means to secure it. . . . But sincerity takes away all the policies of love. The man who can manage his affairs with the care and coolness of his usual habits is not much in earnest. Though the one who feels is less able than the one who does not to take advantage of circumstances as they occur, still I would not change the honourable consciousness of earnest affection and sincerity for the cool caution and procedure of the mind at ease, though the first were doomed to failure and the last were blessed with success.' (hum!!!) [Note evidently added at a later date.]

The last evening they drove to Manston. 'I could not have imagined a ride so pleasant as the one of this evening. . . . The time of day, the scenery we passed through, and the places we visited, were all calm and composed, and heightened the feelings of tranquil enjoyment and perfect confidence which floated round our hearts. . . . Not a moment's alloy of this evening's happiness occurred: everything was delightful to the last instant of my stay with my companion, because she was so.'

. . . .

On the 7th of August he returned to London, and was at Paternoster Row as soon as possible, and pleased them by letters and accounts from Ramsgate, and then endeavoured to get into the usual routine of life and business again.

On the 8th of August he writes to Miss Barnard:—

'Since the week I have passed with you, every moment offers me fresh proof of the power you have over me. I could not at one time have thought it possible that I, that any man, could have been under the dominion of feelings so undivided and so intense; now I think that no other man can have felt or feel as I do.

. . . .

'If your fears return, tell me, that I may search out antidotes, and doubt not that I shall find them. Read or alter my letters; do anything you please to drive them away. Fly to Mrs. Reid for help, and then thank her from me for it. I shall never indeed, as it is, be able to repay her kindness, but I will try to acknowledge it in attentions and affection *to you her sister*.'

November 29th, he writes:—'Is it a proof that the heart is more true because the mouth more frequently declares it? Is it always found that the most exaggerated and hyperbolical are the truest accounts? or is not, on the contrary, the truth always simple and always plain? I should feel myself debased were I to endeavour to gain your heart by many and glowing descriptions; I should debase your idea in my mind were I for a moment to think you could be affected by them. . . .

'What can I call myself to convey most perfectly my affection and love to you? Can I, or can truth, say more than that for this world I am yours?

'M. FARADAY.'

In December he writes:—

'Royal Institution: Tuesday evening.

'My dear Sarah,—It is astonishing how much the state of the body influences the powers of the mind. I have been thinking all the morning of the very delightful and interesting letter I would send you this evening, and now I am so tired, and yet have so much to do, that my thoughts are quite giddy, and run round your image without any power of themselves to stop and admire it. I want to say a thousand kind and, believe me, heartfelt things to you, but am not master of words fit for the purpose; and still, as I ponder and think on you, chlorides, trials, oil, Davy, steel, miscellanea, mercury, and fifty other professional fancies swim before and drive me further and further into the quandary of stupidness.

'From your affectionate

'MICHAEL.'

The four following letters belong to the next year, but they are placed here to complete the account of his marriage.

FARADAY TO MISS SARAH BARNARD.

February 12, 1821.

He writes :—'Do you know I felt a little angry with Edward on first reading your letter, not because he had in some measure prevented me from seeing you this evening, but because, from what you say, he seems to have been a little vexed with you for something arising out of your affection for me; and, as that is a thing which above all others that I possess I value most, so it is one which, though touched in the slightest manner, would soonest put me in a blaze. What! a feeling so kind, so merciful, so good, so disinterested, can it give rise to anything wrong? I shall expect that Edward and all others will take it for granted, even against their own reasoning, that whatever that feeling suggests to you will be right and proper. I must have respect paid to it greater than is paid to myself. All who play with or neglect it, venture that play or disrespect to me on a point upon which, least of all others, I am at all tractable. . . . .

'May every blessing attend you, and, above all, that of a happy mind.

'From your devoted

'M. FARADAY.'

Later he writes :—

FARADAY TO MISS SARAH BARNARD.

'I tied up the enclosed key with my books last night, and make haste to return it lest its absence should occasion confusion. If it has, it will perhaps remind you of the disorder I must be in here also for the want of a key—I mean the one to my heart. However, I know where my key is, and hope soon to have it here, and then the Institution will be all right again. Let no one oppose my gaining possession of it when unavoidable obstacles are removed.

'Ever, my dear girl, one who is perfectly yours,
'M. FARADAY.'

March 11, 1821, Sir H. Davy wrote:—'Dear Mr. Faraday, I have spoken to Lord Spencer, and I am in hopes that your wishes may be gratified; but do not mention the subjec till I see you.' This wish was probably to bring his wife to the Institution. In May he was appointed superintendent of the house and laboratory.

All obstacles were removed, and they were married on June 12. Faraday, desiring that the day should be considered 'just like any other day,' offended some of his near relations by not asking them to his wedding.

In a letter to Mrs. Reid, previous to the marriage, he says, 'There will be no bustle, no noise, no hurry occasioned even in one day's proceeding. In externals, that day will pass like all others, for it is in the heart that we expect and look for pleasure.'

Twenty-eight years after, in the notes of his own life, he wrote:—'On June 12, 1821, he married—an event which more than any other contributed to his earthly happiness and healthful state of mind. The union has continued for twenty-eight years, and has nowise changed, except in the depth and strength of its character.'

Many more letters were written to Miss Barnard during the time of his engagement, but enough are here

1821. ÆT. 29.

given to show what his character was; and this view of him must be brought to a close by the following letter to the unmarried sister of Miss Barnard.

FARADAY TO MISS JANE BARNARD.

'Royal Institution: May 5, 1821.

'My dear Jane,—I know of no circumstance in my life that has contributed, or promises to contribute, so much to my happiness as my acquaintance at your house. In addition to the pleasures that you know of which have become mine, there are others which it has produced that none but myself can feel or understand. Among those which are evident is the possession of your kind good-will and sisterly affection, for though I may flatter myself that it is greater than it really is, yet I hope and believe you will not refuse it me entirely in return for mine. Your sister has managed to open my heart, and set the springs of my affection flowing, when I supposed there had been no source for them; and I shall not be happy unless they embrace and receive a welcome from all that love her. I want to be truly one of your family, and not the separator of Sarah from it. Receive, therefore, this little gift from me as from a brother, and in receiving it let it be as a sister. The pleasure which I shall feel on its acceptance will be greater than any it can cause you, and will be still increased the more readily it is received. So that you observe, I shall every way be still your obliged, and let me add by anticipation

'Your affectionate brother,

'M. FARADAY.

'Miss Jane Barnard, with a gift of a work-box.'

SIR H. DAVY TO FARADAY, ON HIS MARRIAGE.

'Northampton: July 18, 1821.

'My dear Faraday,—You will find, by my troubling you immediately, that I do not consider your kind offer of doing anything for me in my absence merely complimentary.

. . . . . .

'I hope you will continue quite well, and do much during the summer; and I wish you in your new state all that happiness which I am sure you deserve.

'I am, my dear Mr. Faraday, your sincere friend,

'H. DAVY.'

In 1820 the progress of his knowledge in science makes itself evident in the papers he published, and in the letters which he wrote. His first paper was read to the Royal Society on two new compounds of chlorine and carbon, and on a new compound of iodine, carbon, and hydrogen; and with Mr. Stodart, the surgical instrument maker, he published, in the 'Quarterly Journal of Science,' experiments on the alloys of steel, made with a view to its improvement.

A good account of this work is given in his letters to M. de la Rive. His second letter is an abstract of the paper that was published on steel. In it he tells of the artificial formation of Indian steel or wootz; of the alloys of steel with rhodium, silver, platinum, nickel, &c.

Green, Picksley & Co., in Sheffield, for a time used the alloy of steel and silver for fenders, &c., and the alloys of steel and rhodium, iridium and silver, they

1820. ÆT. 28.

made into razors: but this long and difficult piece of work proved of no lasting use.

In the eighth and ninth volumes of the 'Quarterly Journal' he had five papers—one of these was on the decomposition of chloride of silver by hydrogen, and another on a description of a new apparatus for the combustion of the diamond.

In his commonplace-book there is a plan of 'Lessons in Chemistry,' 'Processes for Manipulation.' This was the germ of the book which he afterwards wrote on chemical manipulation, and probably it had its origin in the lessons he gave to his laboratory pupil. Another entry in his commonplace-book is on 'Lecture Subjects:' 1. Application of statics to chemistry. 2. Approximation of mechanical and chemical philosophy. 3. Application of mathematics to actual service and use in the arts. 4. Series of mechanical arts, taning, &c.

He wrote from the Royal Institution his second letter to M. de la Rive, Professor of Chemistry, Geneva.

'April 20, 1820.

'Dear Sir,—I never in my life felt such difficulty in answering a letter as I do at this moment your very kind one of last year. I was delighted, on receiving it, to find that you had honoured me with any of your thoughts, and that you would permit me to correspond with you by letter. But I fear that my intention of meriting that honour has already made me unworthy of it, for whilst waiting continually for any scientific news that might arrive to send you, I have delayed my answer so long as almost to forfeit the right of permission to send one at all. I hope you will attribute my tardiness to its right motive, diffidence of my worthiness to write to you, and that it will not injure me in your estimation. I will promise, if you still grant me the liberty of correspondence, never to err so again.

'I am the more ashamed of my neglect because it is a neglect of gratitude as well as of respect. I am deeply indebted to you for your kind expressions respecting my paper on the sonorous tubes, and its value is very much increased with me by your praise. I regret, however, on the same subject, you should imagine that I thought but little of your experiment with mercury. I made it immediately, and was very much surprised by it, and I only refrained from noticing it because I was afraid of myself, and thought I should apply it wrongly, and thus intrude on your subject without any right or reason. Indeed, I had hopes that you would take up the subject again, and after reviewing the various sonorous phenomena of different kinds, as produced in different ways, would undertake what I had not ventured to do; namely, to draw general conclusions, and develop the laws to which they (the phenomena) were obedient.

'You have honoured me by many questions, and no regret can be greater than mine that I have suffered time to answer them rather than myself. In every line of your kind letter I find cause to reproach myself for delay. The next I will answer more readily, and the fear will be that I shall trouble you too often rather than too seldom.

'You honour me by asking for scientific news, and for any little information of my own. I am sorry that both sources are very barren at present, but *I do hope* that both will improve. Mr. Stodart and myself have lately been engaged in a long series of experiments and trials on steel, with the hope of improving it, and I think we shall in some degree succeed. We are still very much engaged on the subject; but if you will give me leave, I will, when the experiments are more complete, which I expect will be shortly, give you a few notes on them. I succeeded by accident, a few weeks ago, in making artificial plumbago, but not in useful masses. I had heated iron with charcoal dust two or three times over, and in that way got a dark grey, very crystalline, carbonate of iron, of, I believe, a definite composition; but the outside of the button, which had been long in close contact with the charcoal, was converted into excellent plumbago. Since then I have observed among the casters of iron, that when they cast on a facing of charcoal dust, as is the case in fine work, that the surface of the casting is frequently covered with a thin film of plumbago, evidently formed in the same way as in the above experiment.

'We have lately had some important trials for oil in this metropolis, in which I, with others, have been engaged. They have given occasion for many experiments on oil, and the discovery of some new and curious results. One of the trials only is finished, and there are four or five more to come. As soon as I can get time, it is my intention to trace more closely what takes place in oil by heat; and I hope to bribe you to continue to me the honour and pleasure of your correspondence, by saying, that if anything important turns up, I will make it the matter of a letter.

'I am, my dear Sir, with the highest respect, your obliged and humble

'M. FARADAY.

He sends in his third letter to Professor de la Rive an abstract of his paper on steel :—

'Royal Institution : June 26, 1820.

'My dear Sir,—Not long since I troubled you with a letter in which I said I would shortly send you an account of some experiments on steel, made by Mr. Stodart and myself. A paper will appear in the next number of our Journal, which will contain all we have as yet ascertained on the subject; and as the results seem to me to be interesting, I hope you will not be sorry that I keep my promise by mentioning the principal of them to you. In the small way in which only we have as yet worked they are good, and I hope that no failure will occur when the processes are transferred to the manufactory.

'It is possible you may have observed an analysis of wootz, or the Indian steel, published in one of our Journals some time since. I could at that time find nothing in the steel, besides the iron and carbon, but a small portion of the earths, or, as I presume, their metallic bases. On the strength of this analysis, we endeavoured to demonstrate the particular nature of wootz synthetically by combining steel with these metallic bases, and we succeeded in getting alloys which when worked were declared by Mr. Stodart to be equal in all qualities to the best Bombay wootz. This corroboration of the nature of wootz received still stronger confirmation from a property possessed by the alloy in common with wootz; namely, their power of yielding damasked surfaces by the action of acids. When wootz is fused and forged it still retains so much of the crystalline structure as to exhibit, when acted on by very weak sulphuric acid, for some time a beautiful damasked surface. This we have never yet seen produced by pure steel, but it is produced in our imitation of wootz, or alloys of steel with the metal of alumine.

'I must not forget to tell you how we formed our alloys. Many attempts failed: the following method succeeds:—Fuse iron in small pieces with charcoal powder. If the button produced is malleable break it up, and re-fuse it with more charcoal. In this way a carburet of iron will be formed which has its place between steel and plumbago. It is fusible, when broken has a dark grey colour, and is very highly crystalline. It is so brittle that small pieces of it may be rubbed to powder in a mortar. Some of this powdered carburet was then mixed with pure alumine, and the whole strongly heated. A portion of the alumine was reduced by the carbon of the carburet, and a compound of iron, aluminum, and carbon was obtained. Then English cast steel being mixed with about 10 per cent. of this alloy, the whole was fused, and our artificial wootz obtained.

'I presume that the properties of wootz are so well known to you, that I need not stop to say what are the supposed improvements in steel when it is converted into wootz.

'Whilst making the carburet above mentioned, we also succeeded in forming plumbago; but I am afraid this artificial production of it will not be very useful in its application. If iron be heated highly, and long enough, in contact with charcoal, plumbago is always formed. I have some buttons of metal here, weighing two or three ounces, that appear to be solid plumbago. The appearance, however, is deceitful, for it is only on the surface, and to the depth perhaps of $\frac{1}{40}$th of an inch, that the plumbago has been formed. The internal part is composed of the crystalline carburet before mentioned. What is plumbago is very good, and marks excellently well; and though we have never yet been able to fuse powdered plumbago into a mass, yet I think, if it were required to form it in a compact state to work up into pencils, it might be done by imbedding plates of iron about $\frac{1}{20}$th of an inch thick in charcoal, and heating intensely for a long time. This we have not yet had time to try, but intend to do so.

'You will readily suppose that, during nearly two years that we have been at work on this subject, a great deal of useless matter, except as furnishing experience, has accumulated. All this you will rather wish away, so that I shall pass over unimportant alloys, to write of those which promise good results.

'Perhaps the very best alloy we yet have made is that with rhodium. Dr. Wollaston furnished us with the metal, so that you will have no doubts of its purity and identity. One-and-a-half per cent. of it was added to steel, and the button worked. It was very malleable, but much harder than common steel, and made excellent instruments. In tempering the instruments they require to be heated full 70° F. higher than is necessary for the best cast steel, and from this we hope it will possess greater degrees of hardness and toughness. Razors made from the alloy cut admirably.

'Next to the alloy of rhodium comes that of silver, about which there are many curious circumstances. Silver refuses to combine with steel except in very small proportions, and this want of affinity is much greater when the metals are cold than when hot. If, for instance, a hundred parts of steel and one of silver be fused together, cooled, hammered, &c. &c., and then laid in weak sulphuric acid for ten or twelve hours, its structure will be developed, and it will be found to be a congeries of fibres of steel and silver, the one distinct from the other, but intimately mixed in every part. Now, the perfect dispersion of the silver throughout all parts proves that it has been taken up by the steel whilst in fusion; but its separate state of existence shows that it has been rejected from the alloy as it solidified. Indeed, this refusal of the silver by the steel as it cools is so remarkable, that if the hot alloy be observed, globules of silver may be seen extruded from the surface as the temperature falls.

'But, however, we went on diminishing the quantity of silver as long as its separate existence could be observed in the alloys; and when we arrived to a $\frac{1}{500}$th part we found that the whole remained in combination with the steel. This alloy was excellent, all the cutting instruments made of it were of the best quality, and the metal worked without break or flaw, and with remarkable toughness and malleability under the hammer. The alloy of steel and platinum is not so marked by an acquired superiority as the two I have already mentioned, and yet platinum in quantities from one to three per cent. does seem to be of advantage to steel: but we are now continuing this subject. The powerful affinity with which platinum combines with the metals generally, meets with no exception when tried with iron or steel. They unite in all proportions we have tried, from 1 platinum to 100 steel, up to 90 platinum to 20 steel. We expect a good deal from some of these higher compounds.

'I think the affinities of platinum and silver for steel are worth comparing together, though they stand almost together in an electrical arrangement of the metals, and both of them very far from iron or steel: still they do not exhibit attractions for steel at all comparable. Platinum will combine in any proportion, apparently, with steel or iron, and at temperatures so low that the two metals may be welded together at heats barely sufficient to weld iron; whereas it is with difficulty that a $\frac{1}{500}$th part of silver is made in any way to combine with steel.

'I hope, my dear Sir, I have not tired you yet, for I am now going to begin writing across; but I will promise not to detain you very much longer, either by excuses or details.

'We have been induced by the popular idea that meteoric iron would not rust to try the effect of nickel on steel and iron. We made alloys of iron and nickel, varying the latter metal from three to ten per cent., and we thought we found that they were not quite so oxidable as iron alone when exposed with it in green-houses and in our laboratory. But nickel alloyed with steel gave us no hopes. It appeared more oxidable than simple steel, and this fault was not compensated for by any other good quality. So for the present we have dismissed that metal from our experiments, though I expect, as we go on, we shall find many occasions to resume thoughts and intentions which we may have laid down.

'Mr. Children has obliged us with an accurate analysis of the Siberian meteoric iron, and he finds it to contain a very large proportion of nickel. In the mean of three experiments it amounts to 8·96 per cent.

1820. ÆT. 28.

'You cannot imagine how much we have been plagued to get a crucible that will bear the heat we require and use in our experiments. Hessian, Cornish, pipeclay crucibles all fuse in a few minutes, if put into the furnace singly; and our only resource is to lute two or three, one within another, together, so that the whole may not fuse before our alloy has had time to form in the centre. I have seen Hessian crucibles become so soft that the weight of 500 grains of metal has made them swell out like a purse, and the upper part has fallen together in folds like a piece of soft linen; and where three have been put together, the two outer ones have, in less than half an hour, melted off, and flown down into the grate below.

'From these circumstances you will judge of the heat we get. And now I will mention to you an effect which we obtain, and one we can't obtain, both of which a little surprised us. The positive effect is the volatilisation of silver. We often have it in our experiments sublimed into the upper part of the crucible, and forming a fine dew on the sides and cover; so that I have no doubt at present on the volatility of silver, though I had before. The non-effect is the non-reduction of titanium. We have tortured menachanite, pure oxide of titanium, the carbonate, &c., in many ways in our furnace, but have never yet been able to reduce it—not even in combination with iron; and I must confess that now I am very sceptical whether it has ever been reduced at all in the pure state.

'Now I think I have noticed the most interesting points at which we have arrived. Pray pity us that, after two years' experiments, we have got no further; but I am sure, if you knew the labour of the experiments, you would applaud us for our perseverance at least. We are still encouraged to go on, and think that the experience we have gained will shorten our future labours; and if you find the contents of this well-covered sheet of paper interesting, I shall at some future time do myself the honour and pleasure of sending a continuation of it.

'If you should think any of our results worth notice in the 'Bibliothèque,' this letter is free to be used in any way you please. Pardon my vanity for supposing anything I can assist in doing can be worth attention; but you know we live in the good opinion of ourselves and of others, and therefore naturally think better of our own productions than they deserve.

'I am, my dear Sir, very truly and sincerely your obliged 'M. FARADAY.'

Before beginning the account of the higher scientific education which Faraday went through in 1821 two events which had great influence upon his future life must be mentioned. He became a Sandemanian, and he had to endure a false charge of dishonest conduct.

A month after his marriage he made his confession of sin and profession of faith before the Sandemanian Church.

His faith in Christ he considered to be the effect of Divine power—the unmerited gift of God to one who had nothing in him that could be pleasing in His sight. The sense of his own unworthiness, and incapability of doing what was good before God, extended even to this act of professing the truth.

When his wife asked him why he had not told her what he was about to do, he only replied, 'That is between me and my God.' A friend writes:

'When he entered the meeting-house he left his science behind, and he would listen to the prayer and exhortation of the most illiterate brother of his sect with an attention which showed how he loved the word of truth, from whomsoever it came.'

In his lecture on Mental Education, in 1854, he uses the following words:—

'High as man is placed above the creatures around him, there is a higher and far more exalted position within his view; and the ways are infinite in which he occupies his thoughts about the fears, or hopes, or expectations of a future life. I believe that the truth of that future cannot be brought to his knowledge by any exertion of his mental powers, however exalted they may be; that it is made known to him by other teaching than his own, and is received through simple belief of the testimony given. Let no one suppose for a moment that the self-education I am about to commend, in respect of the things of this life, extends to any considerations of the hope set before us, as if man by reasoning could find out God. It would be improper here to enter upon this subject further than to claim an absolute distinction between religious and ordinary belief. I shall be reproached with the weakness of refusing to apply those mental operations which I think good in respect of high things to the very highest. I am content to bear the reproach; yet, even in earthly matters, I believe that the invisible things of Him from the creation of the world are clearly seen, being understood by the things that are made, even His eternal power and Godhead; and I have never seen anything incompatible between those things of man which are within him, and those higher things concerning his future, which he cannot know by that spirit.'

The false charge of dishonesty arose thus :—

Dr. Wollaston was the first person who entertained the possibility of electro-magnetic rotation. He perceived that there was a power not directed to or from the wire in the voltaic circuit, but acting circumferentially round its axis, and upon that he founded his expectations of making the wire revolve on itself. In April, 1821, he came with Sir H. Davy to the laboratory of the Institution to make an experiment. Faraday was not there at the time; but coming in afterwards, he heard the conversation, and this expectation of making a wire in the voltaic circuit revolve on its own axis. He had before this heard a rumour of a wager that Dr. Wollaston would succeed in doing this.

In July, August, and September, Faraday wrote, at the request of the editor, R. Phillips, an historical sketch of electro-magnetism for the 'Annals of Philosophy.' He repeated almost all the experiments he described.

This led him, in the beginning of September, to discover the rotation of a wire in the voltaic circuit round a magnet, and of a magnet round the wire. He could not make the wire and the magnet revolve on their own axis. 'There was not the slightest indication that such was the case.' 'I did not realise Dr. Wollaston's expectation of the rotation of the electro-magnetic wire round its axis; that fact was discovered by M. Ampère at a later date,' he said.

Before he published his paper on these 'new electromagnetical motions,' he tried to see Dr. Wollaston: his object was to ask permission to refer to Dr. Wollaston's views and experiments. He was out of town, and 'by an error of judgment the paper was published without any allusion to his opinions and intentions.' Immediately afterwards, Faraday heard rumours 'affecting his honour and honesty.' He at once asked his friend Mr. Stodart to procure him an interview with Wollaston.

FARADAY TO MR. STODART.

'Royal Institution: Monday, October 8, 1821.

'My dear Sir,—I hear every day more and more of those sounds, which, though only whispers to me, are I suspect spoken aloud amongst scientific men, and which, as they in part affect my honour and honesty, I am anxious to do away with, or at least to prove erroneous in those parts which are dishonourable to me. You know perfectly well what distress the very unexpected reception of my paper on magnetism in public has caused me, and you will not therefore be surprised at my anxiety to get out of it, though I give trouble to you and other of my friends in doing so.

'If I understand aright, I am charged, (1) with not acknowledging the information I received in assisting Sir H. Davy in his experiments on this subject; (2) with concealing the theory and views of Dr. Wollaston; (3) with taking the subject whilst Dr. Wollaston was at work on it; and (4) with dishonourably taking Dr. Wollaston's thoughts, and pursuing them, without acknowledgment, to the results I have brought out.

'There is something degrading about the whole of these charges; and were the *last* of them true, I feel that I could not remain on the terms I now stand at with you or any scientific person. Nor can I, indeed, bear to remain even suspected of such a thing. My love for scientific reputation is not yet so high as to induce me to obtain it at the expense of honour, and my anxiety to clear away this stigma is such that I do not hesitate to trouble you even beyond what you may be willing to do for me.

'I want you, my dear Sir, to procure me an interview with Dr. Wollaston on his return to town; and I wish for this not only to apologise to him if I have unintentionally done him wrong, but to justify myself from the suspicions that are wrongly raised against me. I feel that Dr. Wollaston is so very far above me that even if he does feel himself wronged, he may not permit himself to think it is of any importance, and may therefore think it unnecessary to allow anything to pass on the subject. But, in that case, I appeal to Dr. Wollaston on my own account. His character and talents have raised him to be a patron and protector of science. All men look to his opinion and judgment with respect. If therefore an impression has gone abroad that I have done him an injustice, surely he will listen to my vindication, if not for his own or even my sake, yet for the sake of that situation in which he stands in the scientific world. I am but a young man, and without a name, and it probably does not matter much to science what becomes of me; but if by any circumstances I am subjected to unjust suspicions, it becomes no one more than him who may be said to preside over the equity of science, to assist in liberating me from them.

'With regard to the first charge, I have spoken to Sir H. Davy, and I hope and believe he is satisfied. I wished to apply to him, but knew not where he was till the paper was printed, and immediately I did know

I sent him a rough copy of it. How much I regret the haste which made me print the paper in the last number of the Journal, is known to Sir H. Davy and to you.

'With regard to the second charge, I have to say that I should have been proud to have put into my paper in a more distinct manner what I knew of Dr. Wollaston's theory and experiments; but that I was afraid to attach to it anything which Dr. Wollaston had not published or authorised. At the same time I must state, that all I knew was what is published in the "Journal of Science," vol. x. p. 363, and that Dr. Wollaston expected to make a wire revolve on its own axis; but I did not see the apparatus of Dr. Wollaston, or the experiment he made at the Royal Institution, or any made elsewhere.

'As to the third charge, I had not the slightest notion that Dr. Wollaston was at work, or intended to work, on the subject. It is now near seven months, I believe, since he was at the Royal Institution making an experiment, and I did not know that he intended to pursue it further. If I had thought so I should never have attempted anything on the subject.

'The fourth charge is not true. I had assisted Sir H. Davy in nearly all his experiments, and thus had my mind prepared for the subject; but the immediate cause of my making the experiments detailed in my paper was the writing of the historical sketch of electromagnetism that has appeared in the last two numbers of the "Annals of Philosophy." It was in verifying the positions that I continually had to make mention of in that sketch, that I was led, as described in the commencement of my paper, to ascertain the revolution of the pole round the wire; and then, and then only, Dr.

Wollaston's theory came to my mind. I should have been proud and happy here to have mentioned Dr. Wollaston's experiment of the rotation of the wire on its own axis (the only experiment I had heard of), but it did not succeed with me, or Dr. Wollaston's theory as stated in our Journal. But Dr. Wollaston had not published or avowed either, and I judged (perhaps wrongly) that I had no right in that case to mention it.

'All I ask is to be liberated from the dishonour unjustly attached to me in these charges. I am anxious to apologise to Dr. Wollaston, in any way that I can, for not having mentioned his theory and experiments, if I may be permitted. I need not again urge reasons why Dr. Wollaston should hear me, or receive into his consideration those circumstances which witness for me in this affair that I have erred innocently. But I hope everything through your kindness. Anxiously waiting to hear from you,

'I am, dear Sir, your very obliged and faithful

'M. FARADAY.'

A few days later Faraday wrote himself directly to Dr. Wollaston.

FARADAY TO DR. WOLLASTON.

'October 30, 1821.

'Sir,—I am urged by strong motives respectfully to request your attention for a few moments. The latter end of last month I wrote a paper on electro-magnetism, which I left in the hands of the printer of the "Quarterly Journal," and went into the country. On returning home the beginning of this month, I heard from two or three quarters that it was considered I had not behaved honourably in that paper; and that the wrong I had done was done to you. I immediately wished and endeavoured to see you, but was prevented by the advice of my friends, and am only now at liberty to pursue the plan I intended to have taken at first.

'If I have done anyone wrong, it was quite unintentional, and the charge of behaving dishonourably is not true. I am bold enough, Sir, to beg the favour of a few minutes' conversation with you on this subject, simply for these reasons—that I can clear myself—that I owe obligations to you—that I respect you—that I am anxious to escape from unfounded impressions against me—and if I have done any wrong that I may apologise for it.

'I do not think, Sir, that you would regret allowing me this privilege; for, satisfied in my own mind of the simplicity and purity of my motives in writing that paper, I feel that I should satisfy you; and you would have the pleasure of freeing me from an embarrassment I do not deserve to lay under. Nevertheless, if for any reasons you do not consider it necessary to permit it, I hope I shall not further have increased any unpleasant feeling towards me in your mind.

'I have very much simplified and diminished in size the rotating apparatus, so as to enclose it in a tube. I should be proud if I may be allowed, as a mark of strong and sincere respect, to present one for your acceptance. I am almost afraid to make this request, not because I know of the slightest reason which renders it improper, but because of the uncertain and indefinite form of the rumours which have come about me. But I trust, Sir, that I shall not injure myself with you by adopting the simplest and most direct means of clearing up a misunderstanding that has arisen against me; but that what I do with sincerity you will receive favourably.

'I am, Sir, with great respect, your obedient, humble servant, 'M. FARADAY.'

WOLLASTON TO FARADAY.

'October 31, or November 1.
'(Must have been about November 1.)

'Sir,—You seem to me to labour under some misapprehension of the strength of my feelings upon the subject to which you allude.

'As to the opinions which others may have of your conduct, that is your concern, not mine; and if you fully acquit yourself of making any incorrect use of the suggestions of others, it seems to me that you have no occasion to concern yourself much about the matter. But if you are desirous of any conversation with me, and could with convenience call to-morrow morning, between ten o'clock and half-past ten, you will be sure to find me.

'Ever your most obedient
'W. H. WOLLASTON.

'I name that hour because I shall have occasion to leave home before eleven.'

Of the result of this interview no record remains, but the false charge seemed for a time to have died away.

Faraday went on with his researches, and on December 21 he succeeded in making a wire, through which a current of voltaic electricity was passing, obey the magnetic poles of the earth in the way it does the poles of the bar-magnet.

1821.
ÆT. 29–30.

Dr. Wollaston came three or four times to the laboratory to see the results.

Nothing more was heard of the charge until in March 1823 Sir H. Davy read a paper to the Royal Society on a new phenomenon of electro-magnetism. At the end he said, 'I cannot with propriety conclude without mentioning a circumstance in the history of the progress of electro-magnetism, which, though well known to many Fellows of this Society, has, I believe, never been made public; namely, that we owe to the sagacity of Dr. Wollaston the first idea of the possibility of the rotation of the electro-magnetic wire round its axis by the approach of a magnet; and I witnessed, early in 1821, an unsuccessful experiment which he made to produce the effect.' The proceedings of this meeting of the Royal Society were reported by Mr. Brayley in the 'Annals of Philosophy.' The report inaccurately[1] confused the rotation of the wire on its own axis with the rotation of the wire round the magnet; and in the last five lines the report said, 'Had not an experiment on the subject made by Dr. W. in the laboratory of the Royal Institution, and witnessed by Sir Humphry, failed *merely through an accident* which happened to the apparatus, *he would have been the discoverer of that phenomenon.*'

'Sir Humphry Davy, when he next adverted to the subject with Faraday, in the laboratory of the Institution, said this "was inaccurate and very unjust," and advised Faraday "to draw up a contradiction which

[1] Fifteen years afterwards, Mr. Brayley wrote to Faraday, 'I am unwilling positively to affirm the accuracy of my report of the paper in the face of Sir H. Davy's immediate denial, but I always had, and have still, the strongest impression that accurate it was.' The paper itself was lost.

the editor should report the next month." This Faraday did in these words :—

'We endeavoured last month to give a full report of the important paper read by Sir H. Davy to the Royal Society on March 5. We beg our readers, however, to cancel the five last lines of that report, which are not merely incorrect but untrue; and, anxious to avoid the commission of an act of injustice to a third person, we wish to refer them forward to the original paper when it shall be published.'

This was submitted to Sir H. Davy 'as a proper correction; he altered it, and it became what he wished it should be, and Faraday agreed to it.' It stands thus in the 'Annals of Philosophy' for May: 'Writing only from memory, we have made two errors, one with respect to the rotation of mercury, &c., the other in the historical paragraph, in the conclusion, which, as we have stated it, is unjust to Mr. Faraday, and does not at all convey the sense of the author. We wish, therefore, to refer our readers forward to the original paper, when it shall be published, for the correction of these mistakes.'

'Thus,' says Faraday afterwards, 'I was unjustly subjected to some degree of annoyance, and the more so because this happened at the very time of the occurrence of the condensation of the gases and its consequence, and during the time that my name was before the Royal Society as a candidate for the Fellowship. I do not believe that anyone willingly was the cause of this state of things, but all seemed confusion, and generally to my disadvantage.'

A month after he had been proposed as Fellow of

the Royal Society in 1823, he wrote thus to Mr. Warburton, the most intimate friend of Dr. Wollaston.

FARADAY TO MR. WARBURTON.

'Royal Institution: May 30, 1823.

'Sir,—I have been anxiously waiting the opportunity you promised me of a conversation with you, and from late circumstances am now still more desirous of it than at the time when I saw you in the committee. I am sure you will not regret the opportunity you will afford for an explanation, for I do not believe there is anything you would ask, *after you have communicated with me*, that I should not be glad to do. I am satisfied that many of the feelings you entertain on the subject in question would be materially altered by granting my request; at the same time, as I know more of your opinions by report than otherwise, I am perhaps not well aware of them. It was only lately that I knew you had any feeling at all on the subject. You would probably find yourself engaged in doing justice to one who cannot help but feel that he has been injured, though he trusts unintentionally. I feel satisfied you are not in possession of all the circumstances of the case, but I am also sure you will not wish willingly to remain ignorant of them. Excuse my earnestness and freedom on this subject, and consider for a moment how much I am interested in it.

'I would have called upon you, but I was not aware of the hour at which it would be convenient and agreeable to you to see me, and I have very little time to spare in the day from my duties here; but if you will appoint a time, I will call on you, or do anything you direct to obtain a meeting.

'I am, Sir, your obedient, humble servant,

'M. FARADAY.'

Faraday made the following notes at the end of his copy of this letter to Warburton.

1823. In relation to Davy's opposition to my election at the Royal Society.

Sir H. Davy angry, May 30.

Phillip's report through Mr. Children, June 5.

Mr. Warburton called first time, June 5 (evening).

I called on Dr. Wollaston, and he not in town, June 9.

I called on Dr. Wollaston, and saw him, June 14.

I called at Sir H. Davy's, and he called on me, June 17.

To counteract the opposition which the circulation of the false charge was causing to him he published an historical statement respecting electro-magnetic rotation. Of this he writes, 'I think it worth while saying, that before this historical statement was produced, it was shown to Dr. Wollaston, altered by him in certain places with a pencil, and then declared by him to be perfectly satisfactory. He gave me leave to tell all persons that he was satisfied with it, and thought it convincing; and the manuscript corrected by him is bound up in the quarto volume of my papers from the "Philosophical Transactions."' This manuscript is now in the library of the Royal Institution.

The most important alteration is where Faraday alludes to the state of his knowledge of Dr. Wollaston's ideas, before he made his discovery, and after he had heard of Dr. Wollaston's expectations: 'I throughout the sketch describe attractive and repulsive powers on each side of the wire; but what I thought to be attraction and repulsion in August 1821, Dr. Wollaston long before perceived to be an impulse in one direction only, and upon that knowledge founded his expectations.' This Dr. Wollaston altered to 'perceived to be a power not directed to or from the wire, but acting circumferentially round its axis; and upon that knowledge founded his expectations.' This historical statement was published July 1, 1823, and in it he says:—

'The paper which I first published was written, and the experiments all made, in the beginning of September 1821: it was published on the 1st of October. A second paper was published in the same volume on the last day of the same year. I have been asked why in those papers I made no reference to Dr. Wollaston's opinions and intentions, inasmuch as I always acknowledged the relation between them and my own experiments. To this I answer, that upon obtaining the results described in the first paper, and which I showed very readily to all my friends, I went to Dr. Wollaston's house to communicate them also to him, and to ask permission to refer to his views and experiments. Dr. Wollaston was not in town, nor did he return whilst I remained in town; and as I did not think that I had any right to refer to views not published, and, as far as I knew, not pursued, my paper was printed, and appeared without that reference, whilst I remained in the country. I have regretted ever since I did not delay the publication, that I might have shown it first to Dr. Wollaston.

'Pursuing this subject (at the end of the year), I obtained some other results which seemed to me worthy of being known. Previous to their arrangement in the form in which they appear, I waited on Dr. Wollaston, who was so kind as to honour me with his presence two or three times, and witness the results. My object was then to ask his permission to refer to his views and experiments in the paper which I should immediately publish, in correction of the error of judgment in not having done so before. The impression that has remained on my mind ever since (one-and-twenty months), and which I have constantly expressed to everyone when talking on the subject, is that he wished me not to do so. Dr. Wollaston has lately told me that he cannot recollect the words he used at the time; that as regarded himself his feelings were it should not be done, as regarded me that it should, but that he did not tell me so. I can only say that my memory at this time holds most tenaciously the following words, "I would rather you should not;" but I must of course have been mistaken. However, that is the only cause why the above statement was not made in December 1821.'

On July 8 Mr. Warburton wrote to him.

WARBURTON TO FARADAY.

'Sir,—I have read the article in the "Royal Institution Journal" (vol. xv. p. 288) on electro-magnetic rotation; and without meaning to convey to you that I approve of it unreservedly, I beg to say that, upon the whole, it satisfies me, as I think it will Dr. Wollaston's other friends.

'Having everywhere admitted and maintained that on the score of scientific merit you were entitled to a place in the Royal Society, I never cared to prevent your election, nor should I have taken any pains to form a party in private to oppose you. What I should have done would have been to take the opportunity which the proposing to ballot for you would have afforded me to make remarks in public on that part of your conduct to which I objected. Of this I made no secret, having intimated my intention to some of those from whom I knew you would hear of it, and to the President himself.

'When I meet with any of those in whose presence such conversation may have passed, I shall state that my objections to you as a Fellow are and ought to be withdrawn, and that I now wish to forward your election.

'I am, Sir, your faithful servant,

'HENRY WARBURTON.'

Faraday was at this time out of town: on his return, at the end of August, he wrote to Mr. Warburton.

FARADAY TO WARBURTON.

'Royal Institution: August 29, 1823.

'Sir,—I beg to apologise for not having sooner acknowledged the receipt of your letter: my absence from town will, I hope, plead my excuse.

'I thank you sincerely for your kindness in letting me know your opinion of the statement. Though your approbation of it is not unreserved, yet it very far surpasses what I expected; and I rejoice that you do not now think me destitute of those moral feelings which you remarked to me were necessary in a Fellow of the Royal Society.

'Conscious of my own feelings and the rectitude of my intentions, I never hesitated in asserting my claims or in pursuing that line of conduct which appeared to me to be right. I wrote the statement under this influence, without any regard to the probable result, and I am glad that a step, which I supposed would rather tend to aggravate feelings against me, has, on the contrary, been the means of satisfying the minds of many, and of making them my friends.

'Two months ago I had made up my mind to be rejected by the Royal Society as a Fellow, notwithstanding the knowledge I had that many would do me justice; and, in the then state of my mind, rejection or reception would have been equally indifferent to me. Now that I have experienccd so fully the kindness and liberality of Dr. Wollaston, which has been constant throughout the whole of this affair, and that I find an expression of good-will strong and general towards me, I am delighted by the hope I have of being honoured by Fellowship with the Society; and I thank you sincerely for your promise of support in my election, because I know you would not give it unless you sincercly thought me a fit person to be admitted.

'I am, Sir, your obliged and obedient servant,

'M. FARADAY.'

The almost unanimous election of Faraday into the Royal Society shows that, in the judgment of those who were best qualified to form an opinion, Faraday had for two years been subject to a false charge. From this Davy and Wollaston ought to have been his foremost defenders.

In 1821, the knowledge he got and the work he did is to be seen chiefly in the papers he published, and in the letters which he wrote to Professor de la Rive.

I.

At the end of the previous year he sent to the Royal Society a paper on two new compounds of chlorine and carbon; and on a new substance containing iodine, carbon, and hydrogen. And in July this year he and Phillips sent in another paper, on a new compound of chlorine and carbon. Both papers form part of the 'Philosophical Transactions' for this year.

In the 'Quarterly Journal' he had seven papers. The most important were on the vapour of mercury at common temperatures; on the dissection of crystals; and on a singular property of boracic acid as regards its action on turmeric paper.

II.

Very few letters written in 1821 remain. There is some interest in reading the two last which he wrote to his friend Abbott: one shows his constant occupation; the other his gentle sympathy; and it brings to a final close a correspondence which is invaluable as a record of the nature and self-education of Michael Faraday.

FARADAY TO DE LA RIVE.

'September 12, 1821.

'My dear Sir,—I was extremely gratified the other day on receiving your very kind letter, and also your beautiful little apparatus. I owe you many thanks for them, and have been using the latter, I hope you will say, with some effect. I have not seen M. Prévost, so have not heard any news of your delightful place, except what your letter contains; but I trust all is well.

'I am much flattered and encouraged to go on by your good opinion of what little things I have been able to do in science, and especially as regards the chlorides of carbon.

. . . . . .

'Sir H. Davy's paper is not yet printed; and I hardly know what it is, for Sir H. left town for the country almost before his ideas were put into order on the subject on which he was working.

'You partly reproach us here with not sufficiently esteeming Ampère's experiments on electro-magnetism. Allow me to extenuate your opinion a little on this point. With regard to the experiments, I hope and trust that due weight is allowed to them; but these you know are few, and theory makes up the great part of what M. Ampère has published, and theory in a great many points unsupported by experiments when they ought to have been adduced. At the same time, M. Ampère's experiments are excellent, and his theory ingenious; and, for myself, I had thought very little about it before your letter came, simply because, being naturally sceptical on philosophical theories, I thought there was a great want of experimental evidence. Since then, however, I have engaged on the subject, and have a paper in our "Institution Journal," which will appear in a week or two, and that will, as it contains experiment, be immediately applied by M. Ampère in support of his theory, much more decidedly than it is by myself. I intend to enclose a copy of it to you with the other, and only want the means of sending it.

'I find all the usual attractions and repulsions of the magnetic needle by the conjunctive wire are deceptions, the motions being not attractions or repulsions, nor the result of any attractive or repulsive forces, but the result of a force in the wire, which, instead of bringing the pole of the needle nearer to or further from the wire, endeavours to make it move round it in a never-ending circle and motion whilst the battery remains in action. I have succeeded not only in showing the existence of this motion theoretically, but experimentally, and have been able to make the wire revolve round a magnetic pole, or a magnetic pole round the wire, at pleasure. The law of revolution, and to which all the other motions of the needle and wire are reducible, is simple and beautiful.

'Conceive a portion of connecting wire north and south, the north end being attached to the positive pole of a battery, the south to the negative. A north magnetic pole would then pass round it continually in the apparent direction of the sun, from east to west above, and from west to east below. Reverse the connections with the battery, and the motion of the pole is reversed; or if the south pole be made to revolve, the motions will be in the opposite directions, as with the north pole.

'If the wire be made to revolve round the pole, the

motions are according to those mentioned. In the apparatus I used there were but two plates, and the directions of the motions were of course the reverse of those with a battery of several pairs of plates, and which are given above. Now I have been able, experimentally, to trace this motion into its various forms as exhibited by Ampère's helices, &c., and in all cases to show that the attractions and repulsions are only appearances due to this circulation of the pole, to show that dissimilar poles repel as well as attract, and that similar poles attract as well as repel, and to make, I think, the analogy between the helix and common bar-magnet far stronger than before. But yet I am by no means decided that there are currents of electricity in the common magnet. I have no doubt that electricity puts the circles of the helix into the same state as those circles are in that may be conceived in the bar magnet, but I am not certain that this state is directly dependent on the electricity, or that it cannot be produced by other agencies; and therefore, until the presence of electrical currents be proved in the magnet by other than magnetical effects, I shall remain in doubt about Ampère's theory.

. . . . . .

'Wishing you all health and happiness, and waiting for news from you,

'I am, my dear Sir, your very obliged and grateful

'M. FARADAY.'

1821.
Æt. 29–30.

FARADAY TO DE LA RIVE.

'Royal Institution: November 16, 1821.

'Dear Sir,—Herewith you will receive copies of my papers which I mentioned in a letter I sent to you, per post, a month or two ago, and which I hope you will do me the favour to accept. I also send in this packet a little apparatus I have made to illustrate the rotatory motion on a small scale. The rod below is soft iron, and consequently can have its inner end made north or south at pleasure by contact of the external end with one of the poles of a magnet. To make the apparatus act, it is to be held upright with the iron pin downwards; the north or south pole of a magnet to be placed in contact with the external end of the iron pin and then the wires of a voltaic combination connected, one with the upper platinum wire, the other with the lower pin or magnet: the wire within will then rotate, if the apparatus is in order, in which state I hope it will reach you. Good contacts are required in these experiments.

'Now let me know what is doing with you, for I long for news from your southern parts. An Italian gentleman, who is on his return home, will give this packet to you; at least, I hope so, for I want it to reach you safe. I am excessively busy, too much so at present to try my hand at anything more, or even to continue this letter many lines further: but I hope soon to have a little news on steel to send you.

'I am, my dear Sir, as ever, your very obliged and faithful

'M. FARADAY.

'A single pair of plates two or three inches square, or four inches square, is quite large enough for the apparatus.'

FARADAY TO ABBOTT.

'Royal Institution: Feb. 1, 1821.

'Dear A——, I read your letter informing us of Robert's safe arrival out, and was very glad to find all was well. I hope all will continue so, and that when he comes home he will be every way pleased with his voyage.

'I should have written to you before, but have not yet been able to get a spare evening to appoint for our meeting. However, I hope that Monday week will be free for me, and will try to keep it so, but I will write you again before this.

'I am, dear A., yours very truly,

'M. FARADAY.'

FARADAY TO ABBOTT (ON THE DEATH OF HIS BROTHER).

'Royal Institution, May 15, 1821.

'My dear A——, The receipt of your letter has distressed and grieved me sadly, and I feel how much you must be overwhelmed by this sudden wave of affliction. I was looking forward to a cheerful joyous return, with health and strength invigorated by the voyage. I would fain hope there is some mistake in the account you have received, but the tone of your letter prevents me, when I refer to it.

'These things come over us so suddenly, and with such overpowering force, that no reasoning or philosophy can bear up against them, and the only duty left that it is possible to exert is resignation. You must bear up, my dear A., and comfort and encourage your father and sister, though you do your own feelings violence in the effort. Tell them, but not so as to reawaken sorrows that may be lulled for a moment, how much I feel on this occasion. And believe me, dear A., your distressed friend,

'M. FARADAY.'

In 1822 the progress of Faraday's higher education is seen, (1) in his published papers; (2) in a new note-book, which he called 'Chemical Notes, Hints, Suggestions, and Objects of Pursuit;' and (3) in his laboratory book.

The contrast between the calm of 1822 and the storm of that which was past and of that which was to come is very remarkable.

In July he took his wife and her mother to Ramsgate, and left them there whilst he went for a fortnight with his friend Richard Phillips to Mr. Vivian's, near Swansea, to try a new process in the copper-works.

During his absence he wrote three letters to Mrs. Faraday, full of deep feeling and gentle energy. These are the only records (4) that remain for this year.

I.

A paper on the alloys of steel, by Stodart and Faraday, was read to the Royal Society, and printed in the 'Transactions.' In the 'Quarterly Journal of Science' he had six papers, on some new electro-magnetical motions, and on the theory of magnetism; description of an electro-magnetical apparatus for the exhibition of rotatory motion, note on new electro-magnetical motions; on the changing of vegetable colours as an alkaline property, and on some bodies possessing it, and on the action of salts on turmeric paper; on hydriodide of carbon; on a new compound of iodine and carbon.

II.

He began a fresh volume, which he called 'Chemical Notes, Hints, Suggestions, and Objects of Pursuit.' To it he transferred many of the queries out of his commonplace-book, but he separated his subjects under different heads. Probably at some later period he added this preface, 'I already owe much to these notes, and think such a collection worth the making by every scientific man. I am sure none would think the trouble lost after a year's experience.' When a query got its answer, he drew his pen through it, and wrote the date of the answer across it. In this book are the first germs, in the fewest possible words, of his future discoveries.

For example:—

'General effects of compression, either in condensing gases or producing solutions, or even giving combinations at low temperatures.'

'Convert magnetism into electricity.'

'Do pith balls diverge by disturbance of electricities in consequence of induction or not?'

'State of electricity in the interior and on the surface of conductors, and on the surface of holes through them.'

'Light through gold leaf on to zinc or most oxidable metals, these being poles—or on magnetic bars.'

'Transparency of metals. Sun's light through gold leaf. Two gold leaves made poles—light passed through one to the other.'

III.

The notes in the laboratory book are not of importance. The action of chlorine and nitro-muriatic acid on different substances, as wax, naphthaline, alcohol, led to no result.

One remarkable experiment, which he made on September 10 this year, must be here mentioned, for it bears upon the last experiment he ever made, on March 12, 1862.

For forty years the same subject rose again and again in his mind, and no failure, and no success however great, could make him give up his search after the discovery of the relation of electricity and magnetism to light.

It will be seen that once, and once only, he got all the evidence he wanted of the action of electricity and magnetism upon light; but in the experiment this year, and ever after, with the one great exception, the proof was not forthcoming: and he ends his last experiment by saying, 'Not the slightest effect on the polarised or unpolarised ray was observed.'

The note in the laboratory book runs thus:—

'Polarised a ray of lamp-light by reflection, and endeavoured to ascertain whether any depolarising action (was) exerted on it by water placed between the poles of a voltaic battery in a glass cistern; one Wollaston's trough used; the fluids decomposed were pure water, weak solution of sulphate of soda, and strong sulphuric acid: none of them had any effect on the polarised light, either when out of or in the voltaic circuit, so that no particular arrangement of particles could be ascertained in this way.'

IV.

FARADAY TO MRS. FARADAY.

'Paternoster Row: Sunday evening, July 21, 1822.

'Anxious as I have been to use the only means of communication with you that is left me for the present, yet I have delayed writing till this evening, though I felt certain you would have been rendered happy by a letter from me to-day; but I had left it doubtful whether I should write yesterday, and, when I got home, found many reasons for deferring it, though altogether they were hardly strong enough to counterbalance the single one of giving you pleasure a day earlier. But I must not, my dear girl, suffer my love to you to run away at all times with the prudential reasons which, though small, at various times offer themselves; so I resolved, notwithstanding my fingers tingled to write to you, and you, I knew, would be anxious for my letter, to delay it a day, as well for practice in forbearance as for the convenience.

'I perceive that if I give way to my thoughts, I shall write you a mere love-letter, just as usual, with not a particle of news in it: to prevent which I will constrain myself to a narrative of what has happened since I left you up to the present time, and then indulge my affection. . . In the evening I walked up to the Institution; had a letter from Mr. Brande, which was as well as I expected, and gave me leave to go whenever it was necessary for my health's sake: and then returned home.

'Yesterday was a day of events—little, but pleasant. I went in the morning to the Institution, and in the course of the day analysed the water, and sent an account of it to Mr. Hatchett. Mr. Fisher I did not see. Mr. Lawrence called in, and behaved with his usual generosity. He had called in the early part of the week, and, finding that I should be at the Institution on Saturday only, came up, as I have already said, and insisted on my accepting two ten-pound bank-notes for the information he professed to have obtained from me at various times. Is not this handsome? The money, as you know, could not have been at any time more acceptable; and I cannot see any reason, my dear love, why you and I should not regard it as another proof, among many, that our trust should without a moment's reserve be freely reposed on Him who provideth all things for His people. Have we not many times been reproached, by such mercies as these, for our caring after food and raiment and the things of this world?

'On coming home in the evening, i.e. coming to Paternoster Row home, I learned that Mr. Phillips had seen C., and had told her we should not leave London until Monday evening. So I shall have to-morrow to get things ready in, and I shall have enough to do. I fancy we are going to a large mansion and into high company, so I must take more clothes. Having the 20*l*., I am become bold. . . . .

'And now, how do my dear wife and mother do? Are you comfortable? are you happy? are the lodgings convenient, and Mrs. O. obliging? Has the place done you good? Is the weather fine? Tell me all things as soon as you can. I think if you write directly you get this it will be best, but let it be a long letter. I do not know when I wished so much for a long letter as

I do from you now. You will get this on Tuesday, and any letter from you to me cannot reach Swansea before Thursday or Friday—a sad long time to wait. Direct to me, Post Office, Swansea; or perhaps better, to me at — Vivian Esq., Marino, near Swansea, South Wales. . . . .

'And now, my dear girl, I must set business aside. I am tired of the dull detail of things, and want to talk of love to you; and surely there can be no circumstances under which I can have more right. The theme was a cheerful and delightful one before we were married, but it is doubly so now. I now can speak, not of my own heart only, but of both our hearts. I now speak, not with any doubt of the state of your thoughts, but with the fullest conviction that they answer to my own. All that I can now say warm and animated to you, I know that you would say to me again. The excess of pleasure which I feel in knowing you mine is doubled by the consciousness that you feel equal joy in knowing me yours. Oh, my dear Sarah, poets may strive to describe and artists to delineate the happiness which is felt by two hearts truly and mutually loving each other; but it is beyond their efforts, and beyond the thoughts and conceptions of anyone who has not felt it. I have felt it and do feel it, but neither I nor any other man can describe it; nor is it necessary. We are happy, and our God has blessed us with a thousand causes why we should be so. Adieu for to-night. . . . .

'All here send their love and affections to you both. Mine you can estimate perfectly. I constantly feel as if my love had been increasing continually up to the present moment, and yet could not possibly get stronger: such willing believers are we in the tales told by our passion.

'You shall shortly hear again from your most affectionate and devoted husband,

'M. FARADAY.'

TO MRS. FARADAY.

'Marino, near Swansea: July 25, 1822.

'I have just stolen upstairs into my own room to write to you. I intended to have written this morning, so that the letter should go by to-night's mail, but business at the furnaces detained me till half-past five o'clock; then, as dinner was to take place at six o'clock, and there were great persons to be at it, I was obliged to haste in my dressing, so as not to detain them. After dinner the tedious fashion of remaining at table could not be broken by me alone; so at half-past nine, and not before, we went into the drawing-room to tea: here I was detained half-an-hour, and then stole away to converse a little with you.'

. . . . . .

He then describes his journey and occupations at the works, and ends:—

'I forget myself; I was thinking you were in London. When my thoughts are on you, other things are out of my mind: but I know you will be rather pleased than otherwise with a fault which is the result of the earnest anxious affection of your husband.

'M. FARADAY.'

TO MRS. FARADAY.

'Marino: Sunday, July 28, 1822.

'My dearly beloved Wife,—I have just read your letter again, preparatory to my writing to you, that my thoughts might be still more elevated and quickened than before. I could almost rejoice at my absence from you, if it were only that it has produced such an earnest and warm mark of affection from you as that letter. Tears of joy and delight fell from my eyes on its perusal. I think it was last Sunday evening, about this time, that I wrote to you from London; and I again resort to this affectionate conversation with you, to tell you what has happened since the letter which I got franked from this place to you on Thursday I believe.

'You can hardly imagine how rejoiced I was to get your letter. You will have found by this time how much I expected it, and it came almost to a moment. As soon as I entered the breakfast-room Mr. Vivian gave it to me. Blessings on you, my girl; and thanks, a thousand thanks to you for it.

'We have been working very hard here at the copper works, and with some success. Our days have gone on just as before. A walk before breakfast; then breakfast; then to the works till four or five o'clock, and then home to dress, and dinner. After dinner, tea and conversation. I have felt doubly at a loss to-day, being absent from both the meeting and you. When away from London before, I have had you with me, and we could read and talk and walk; to-day I have had no one to fill your place, so I will tell you how I have done. There are so many here, and their dinner so late and long, that I made up my mind to avoid it, though, if possible, without appearing singular. So, having remained in my room till breakfast time, we all breakfasted together, and soon after Mr. Phillips and myself took a walk out to the Mumbles Point, at the extremity of this side of the bay. There we sat down to admire the beautiful scenery around us, and, after we had viewed it long enough, returned slowly home. We stopped at a little village in our way, called Oystermouth, and dined at a small, neat, homely house about one o'clock. We then came back to Marino, and after a little while again went out—Mr. Phillips to a relation in the town, and myself for a walk on the sands and the edge of the bay. I took tea in a little cottage, and, returning home about seven o'clock, found them engaged at dinner, so came up to my own room, and shall not see them again to-night. I went down for a light just now, and heard them playing some sacred music in the drawing-room; they have all been to church to-day, and are what are called regular people.

'The trial at Hereford is put off for the present, but yet we shall not be able to be in town before the end of this week. Though I long to see you, I do not know when it will be; but this I know, that I am getting daily more anxious about you. Mr. Phillips wrote home to Mrs. Phillips from here even before I did—i.e. last Wednesday. This morning he received a letter from Mrs. Phillips (who is very well) desiring him to ask me for a copy of one of my letters to you, that he may learn to write love-letters of sufficient length. He laughs at the scolding, and says it does not hurt at a distance.

'Mr. Vivian has just been up to me. They had

missed me and did not understand it. He wished me to go down to tea, or, at least, to send some up, both of which I declined. It is now ten o'clock, and he has just left me. He has put the train of my thoughts all in confusion. I want to know when Jane comes home, and who has the kindness to visit you. It seems to me so long since I left you that there must have been time for a great many things to have happened. I expect to see you with such joy when I come home that I shall hardly know what to do with myself. I hope you will be well and blooming, and animated and happy, when you see me. I do not know how we shall contrive to get away from here. We certainly shall not have concluded before Thursday evening, but I think we shall endeavour earnestly to leave this place on Friday night, in which case we shall get home late on Saturday night. If we cannot do that, as I should not like to be travelling all day on Sunday, we shall probably not leave until Sunday night; but I think the first plan will be adopted, and that you will not have time to answer this letter. I expect, nevertheless, an answer to my last letter—i.e. I expect that my dear wife will think of me again. Expect here means nothing more than I trust and have a full confidence that it will be so. My kind girl is so affectionate that she would not think a dozen letters too much for me if there were time to send them, which I am glad there is not.

'Give my love to our mothers as earnestly as you would your own, and also to Charlotte or John, or any such one that you may have with you. I have not written to Paternoster Row yet, but I am going to write now, so that I may be permitted to finish this letter

1823. ÆT. 31.

here. I do not feel quite sure, indeed, that the permission to leave off is not as necessary from my own heart as from yours.

'With the utmost affection—with perhaps too much—I am, my dear wife, my Sarah, your devoted husband,

'M. FARADAY.'

In 1823 his progress in scientific knowledge appears, (1) in his publications and notes, and (2) in the letters which he wrote to Prof. de la Rive and Mr. Huxtable; (3) the fresh outburst of the storm that began in 1821 showed his character; and (4) the scientific honours which he now began to receive are evidence of his reputation.

I.

He had two papers published in the 'Philosophical Transactions'—the first on fluid chlorine, and the second on the condensation of several gases into liquids. He had eleven papers in the 'Quarterly Journal.' The chief were on the temperature produced by vapour; historical statement respecting electro-magnetic rotation; change of musket balls in Shrapnell shells; on the action of gunpowder on lead; on the purple tint of plate glass affected by light; and historical statement respecting the liquefaction of gases.

In his laboratory book there is a note, dated June 26, regarding the diffusion of gases, which is of some interest. In 1817 and 1819 Faraday had published experiments on the passage of different gases through capillary tubes, &c.; now he writes, 'made a mixture of one vol. oxygen and two vols. hydrogen, filled five dry bottles over mercury with the mixture, and also four bottles over water; left them in glasses inverted over mercury and water.' 'Some of the bottles were put in the sun's rays and daylight, and some in a dark place.' In July 1824 they were examined, and he drew the conclusions: 'that mercury cannot confine gases perfectly, and that no contraction took place in the dark, nor (most probably) in the daylight either.' It will be seen that this result was doubted in 1825, but not by Faraday.

II.

He gave to Professor de la Rive and Mr. Huxtable an account of his experiments on the condensation of gases. It will be seen that in the course of these researches he was exposed to much danger.

TO PROFESSOR DE LA RIVE.

'Royal Institution: March 24, 1823.

'My dear Sir,—Though it is now some time since I wrote to you, yet the event connected with it is so fresh in my mind that it seems but a week or two ago. Dr. Marcet called on me, not much more than a week before his death, to say how glad he would be to take any parcel or letter in charge for you; and I, accordingly, wrote a letter, and put together such copies of my papers as I had by me and which you had not received, that you might have them at his hands. Alas THE EVENT! (Dr. Marcet died a few days afterwards.)

'I do not know whether you have received or are likely to receive these things from the persons into whose care Dr. Marcet's papers fell. I hope you will, for I have not other copies of them, and I am anxious they should be honoured by being placed in your hands. But I thought I would write you by the post rather than not write at all. I wish and beg to express my best acknowledgments to M. de la Rive, your son, who has honoured me with a copy of his excellent memoir. I hope for the sake of this new branch of science that he is pursuing it. That which he has done proves what he may do. I hope you will do me the kindness to speak of me to him in the best way you can, for I am always anxious to obtain the good-will and commendation of those who are themselves worthy of praise.

'I have been at work lately, and obtained results which I hope you will approve of. I have been interrupted twice in the course of experiments by explosions, both in the course of eight days—one burnt my eyes, the other cut them; but I fortunately escaped with slight injury only in both cases, and am now nearly well. During the winter I took the opportunity of examining the hydrate of chlorine, and analysing it; the results, which are not very important, will appear in the next number of the 'Quarterly Journal,' over which I have no influence. Sir H. Davy, on seeing my paper, suggested to me to work with it under pressure, and see what would happen by heat, &c. Accordingly I enclosed it in a glass tube hermetically sealed, heated it, obtained a change in the substance, and a separation into two different fluids; and upon further examination I found that the chlorine and water had separated from each other, and the chlorine gas, not being able to escape, had condensed into the liquid form. To prove that it contained no water, I dried some chlorine gas, introduced it into a long tube, condensed it, and then cooled the tube, and again obtained fluid chlorine. Hence what is called chlorine gas is the vapour of a fluid.

'I have written a paper which has been read to the Royal Society, and to which the president did me the honour to attach a note, pointing out the general application and importance of this mode of producing pressure with regard to the liquefaction of gases. He immediately formed liquid muriatic acid by a similar means; and pursuing the experiments at his request, I have since obtained sulphurous acid, carbonic acid, sulphuretted hydrogen, euchlorine, and nitrous oxide in the fluid state, quite free from water. Some of these require great pressure for this purpose, and I have had many explosions.

'I send you word of these results because I know your anxiety to hear of all that is new; but do not mention them publicly (or at least the latter ones) until you hear of them either through the journals, or by another letter from me or from other persons, because Sir Humphry Davy has promised the results in a paper to the Royal Society for me, and I know he wishes first to have them *read there*: after that they are at your service.

'I expect to be able to reduce many other gases to the liquid form, and promise myself the pleasure of writing you about them. I hope you will honour me with a letter soon.

'I am, dear Sir, very faithfully, your obliged servant,

'M. FARADAY.'

1823.
ÆT. 31.

TO HUXTABLE.

'Royal Institution: March 25, 1823.

'Dear Huxtable,—I met with another explosion on Saturday evening, which has again laid up my eyes. It was from one of my tubes, and was so powerful as to drive the pieces of glass like pistol-shot through a window. However, I am getting better, and expect to see as well as ever in a few days. My eyes were filled with glass at first.

'When you see Magrath, who I hope is improving fast, tell him I intended calling upon him, but my second accident has prevented me.

'Yours ever,
'M. FARADAY.'

III.

On May 1 his certificate as candidate for the fellowship of the Royal Society was read for the first time. It was drawn up by his friend Mr. Richard Phillips.

'Mr. Michael Faraday, a gentleman eminently conversant in chemical science, and author of several papers, which have been published in the "Transactions" of the Royal Society, being desirous of becoming a Fellow thereof, we, whose names are undersigned, do of our personal knowledge recommend him as highly deserving that honour, and likely to become a useful and valuable member.'

Twenty-nine names follow; the first four signatures, obtained by Mr. Phillips, were Wm. H. Wollaston, J. G. Children, Wm. Babington, Sir W. Herschel.

Perhaps Sir H. Davy as president, and Mr. Brande as secretary, were unable to sign this proposal. It is quite certain that some jealousy had sprung up in the mind of Davy, and this year a fresh cause of bad feeling arose. Thirteen years afterwards, Faraday gives an account of this to his friend Richard Phillips. It was published in the 'Philosophical Magazine' for 1836.

FARADAY TO R. PHILLIPS.

'Royal Institution: May 10, 1836.

'My dear Sir,—I have just concluded looking over Dr. Davy's life. . . .

'I regret that Dr. Davy has made that necessary which I did not think before so; but I feel that I cannot, after his observation, indulge my earnest desire to be silent on the matter, without incurring the risk of being charged with something opposed to an *honest* character. This I dare not risk; but in answering for myself, I trust it will be understood that I have been driven unwillingly into utterance.

. . . . . .

'The facts of the case, as far as I know them, are these:—In the spring of 1823 Mr. Brande was Professor of Chemistry, Sir Humphry Davy Honorary Professor of Chemistry, and I Chemical Assistant in the Royal Institution. Having to give personal attendance on both the morning and afternoon chemical lectures, my time was very fully occupied. Whenever any circumstance relieved me in part from the duties of my situation, I used to select a subject of research and try my skill upon it. Chlorine was with me a favourite object, and having before succeeded in discovering new compounds of that element with carbon, I had considered that body more deeply, and resolved to resume its con-

1823.
ÆT. 31.

sideration at the first opportunity. Accordingly, the absence of Sir H. Davy from town having relieved me from a part of the laboratory duty, I took advantage of the leisure and the cold weather, and worked upon frozen chlorine. On Sir H. Davy's return to town, which I think must have been about the end of February or beginning of March, he inquired what I had been doing, and I communicated the results to him as far as I had proceeded, and said I intended to publish them in the "Quarterly Journal of Science."

'It was then that he suggested to me the heating of the crystals in a closed tube, and I proceeded to make the experiment, which Dr. Paris witnessed, and has from his own knowledge described (Paris's "Life of Davy," vol. ii. p. 210).

'I did not at that time know what to anticipate, for Sir H. Davy *had not told me his expectations*, and I had not reasoned so deeply as he appears to have done. Perhaps he left me unacquainted with them to try my ability.

'How I should have proceeded with the chlorine crystals without the suggestion I cannot now say; but with the hint of heating the crystals in a closed tube ended for the time Sir H. Davy's instructions to me, and I puzzled out for myself, in the manner Dr. Paris describes, that the oil I had obtained was condensed chlorine.

. . . . . .

'When my paper was written, it was, according to a custom consequent upon our relative positions, submitted to Sir H. Davy (as were all my papers for the "Philosophical Transactions" up to a much later period), and he altered it as he thought fit. This practice was one of great kindness to me, for various grammatical mistakes and awkward expressions were from time to time thus removed, which might else have remained.

'To this paper Sir H. Davy added a note, in which he says, "In desiring Mr. Faraday to expose the hydrate of chlorine to heat in a closed glass tube, it *occurred to me* that one of three things would happen: that it would become fluid as a hydrate; or that a decomposition of water would occur, and euchlorine and muriatic acid be formed; or that the chlorine would separate in a condensed state." And then he makes the subject his own by condensing muriatic acid, and states that he had "requested me (of course as Chemical Assistant) to pursue these experiments, and to extend them to all the gases which are of considerable density, or to any extent soluble in water," &c. This I did, and when he favoured me by requesting that I would write a paper on the results, I began it by stating that Sir H Davy did me the honour to request I would continue the experiments, which I have done under his general direction, and the following are some of the results already obtained. And this paper being immediately followed by one on the application of these liquids as mechanical agents by Sir H. Davy, he says in it, "one of the principal objects I had in view in *causing experiments to be made* on the condensation of different gaseous bodies by generating them under pressure," &c.

. . . .

'I have never remarked upon or denied Sir H. Davy's right to his share of the condensation of chlorine or the other gases; on the contrary, I think that I long ago did him full "justice" in the papers themselves. How could it be otherwise? He saw and revised the manuscripts; through his hands they went to the Royal Society, of which he was President at the time; and he saw and revised the printer's proofs. Although he did not tell me of his expectations when he suggested the heating the crystals in a closed tube, yet I have no doubt that he had them; and though perhaps I regretted losing my subject, I was too much indebted to him for much previous kindness to think of saying that that was mine which he said was his. But *observe* (for my sake), that Sir H. Davy nowhere states that he told me what he expected, or contradicts the passages in the first paper of mine which describe my course of thought, and in which I claim the development of the actual results.

'All this activity in the condensing of gases was simultaneous with the electro-magnetic affair; and I had learned to be cautious upon points of right and priority. When therefore I discovered, in the course of the same year, that *neither I nor Sir H. Davy* had the merit of first condensing the gases, and especially chlorine, I hastened to perform what I thought right, and had great pleasure in spontaneously doing justice and honour to those who deserved it. (Monge and Clouet had condensed sulphurous acid probably before the year 1800; Northmore condensed chlorine in the years 1805 and 1806—"Nicholson's Journal," xii. xiii.) I therefore published on January 1, in the following year, 1824, a historical statement of the liquefaction of gases ("Quarterly Journal of Science," xvi. 229).

. . . . . .

'The value of this statement of mine has since been

fully proved, for upon Mr. Northmore's complaint, ten years after, with some degree of reason, that great injustice had been done to him in the affair of the condensation of gases, and his censure of "the conduct of Sir H. Davy, Mr. Faraday, and several other philosophers for withholding the name of the first discoverer," I was able by referring to the statement to convince him and his friends, that if my papers had done him wrong, *I at least* had endeavoured also to do him right. ("Philosophical Magazine," 1834, iv. p. 261.)

'Believing that I have now said enough to preserve my own "honest fame" from any injury it might have risked from the mistakes of Dr. Davy, I willingly bring this letter to a close, and trust that I shall never again have to address you on the subject.

'I am, my dear Sir, yours, &c.

'M. FARADAY.'

Note afterwards added:—'BEFORE my account of the hydrate could be printed (April 1823), the other experiments were made, and Davy's note to the Royal Society read (March 13, 1823).'

That Sir H. Davy actively opposed Faraday's election is no less certain than it is sad.

Many years ago, Faraday gave a friend the following facts, which were written down immediately:—'Sir H. Davy told me I must take down my certificate. I replied that I had not put it up; that I could not take it down, as it was put up by my proposers. He then said I must get my proposers to take it down. I answered that I knew they would not do so. Then he said, I as President will take it down. I replied that

I was sure Sir H. Davy would do what he thought was for the good of the Royal Society.'

Faraday also said that one of his proposers told him that Sir H. Davy had walked for an hour round the courtyard of Somerset House, arguing that Faraday ought not to be elected. This was probably about May 30. On June 29, Sir H. Davy ends a note, 'I am, dear Faraday, very sincerely your well-wisher and friend.' So that outwardly the storm rapidly passed away; and when the ballot was taken, after the certificate had been read at ten meetings, there was only one black ball.

In 1835, Faraday writes:—

'I was by no means in the same relation as to scientific communication with Sir Humphry Davy after I became a Fellow of the Royal Society as before that period; but whenever I have ventured to follow in the path which Sir Humphry Davy has trod, I have done so with respect and with the highest admiration of his talents; and nothing gave me more pleasure, in relation to my last published paper, the eighth series (of "Experimental Researches"), than the thought that, whilst I was helping to elucidate a still obscure branch of science, I was able to support the views advanced twenty-eight years ago, and for the first time, by our great philosopher.'

The Athenæum Club was formed in 1823, and Faraday was the first Secretary; but 'finding the occupation incompatible with his pursuits, he resigned in May 1824.' 'The original prospectus and early lists of members have his name attached to them.' His friend Magrath was made Secretary in his place.

IV.

The first scientific honour which was paid to Faraday in England came from the Cambridge Philosophical Society. The reputation which he had gained abroad is seen in his election this year as corresponding member of the Academy of Sciences, the highest scientific society in France; whilst in Italy he was made correspondent of the Accademia dei Georgofili of Florence.

In all, Faraday received not less than ninety-five honorary titles and marks of merit; and to the end of his life he could say what he said to Mr. Spring Rice in 1838, when he was asked why he received a pension. What were his titles? He answered, 'One title, namely, that of F.R.S., was sought and paid for; all the rest were spontaneous offerings of kindness and good-will from the bodies named.'

In 1854 he answers Lord Wrottesley, 'I cannot say I have not valued such distinctions; on the contrary, I esteem them very highly, but I do not think I have ever worked for or sought them.'

In 1824–25 the progress of Faraday is but slightly marked. His scientific life is seen, (1) in the works he published; (2) in the notes he made in the laboratory book; (3) in the lectures he delivered, and in the appointments and titles that were given to him; (4) two letters, one to his wife and the other to his sister, are the only indications of his character at this time.

I.

In 1824 he published five papers in the 'Quarterly Journal of Science.' The next year he had eight papers in that journal. The most important of these were, (1) on some cases of the formation of ammonia, and on the means of testing the presence of minute portions of nitrogen in certain states (whence the ammonia comes in organic substances containing no nitrogen, and in inorganic substances, and in some metals, he was unable to discover—the dust with which air is adulterated was not then suspected); (2) on the substitution of tubes for bottles in the preservation of certain fluids, as chloride of sulphur; (3) on the composition of crystals of sulphate of soda.

His chief scientific work was published in a paper in the 'Philosophical Transactions,' on new compounds of carbon and hydrogen, and on certain other products obtained during the decomposition of oil by heat.

The Portable Gas Company at this period condensed oil-gas with a pressure of thirty atmospheres. A thousand cubical feet of good gas yielded one gallon of a fluid which was a mixture of different compounds of carbon and hydrogen. The most important was the bicarburet of hydrogen, as Faraday named it. It has been since called benzol, and now benzine. It is at present prepared in immense quantities for the manufacture of the aniline colours. He had a paper on the formation of ammonia in Thomson's 'Annals of Philosophy.' In relation to his future work, the following experiments are of interest: they are in the laboratory book for 1824.

October 11th.—Attempt to ascertain polarisation of

crystals. A small perfect crystal of nitre, about two inches long, was suspended in succession by a single silkworm fibre and a spider's thread, each being about fourteen inches long. They were hung from the top of a glass jar as balances of torsion, then a very large crystal of nitre was placed beneath the small crystal, and as near as could be to allow freedom of motion; but after long examination no tendency to direction relatively to the large crystal could be observed.

On December 28, 1824, he records his first experiment on magnetic electricity. It is published in the 'Quarterly Journal of Science,' July 1825, p. 338:—

'As the current of electricity produced by a voltaic battery, when passing through a metallic conductor, powerfully affects a magnet, tending to make its poles pass round a wire, and in this way moving considerable masses of matter, it was supposed that a reaction would be exerted upon the electric current, capable of producing some visible effect; and the expectation being, for various reasons, that the approximation of the pole of a powerful magnet would diminish the current of electricity, the following experiment was made: the poles of a battery of from two to thirty four-inch plates were connected by a metallic wire, formed in one part into a helix with numerous convolutions, whilst into the circuit at another part was introduced a delicate galvanometer. The magnet was then put, in various positions and to different extents, into the helix, and the needle of the galvanometer was noticed: no effect, however, upon it could be observed. The circuit was made very long, short, of wires of different metals and different diameters, down to extreme fineness, but the results were always the same. Magnets more or less powerful were used, some so strong as to bend the wire in its endeavours to pass round them. Hence it appears that, however powerful the action of an electric current may be upon a magnet, the latter has no tendency by reaction to diminish or increase the intensity of the former—a fact which, though of a negative kind, appears to me to be of some importance.'

II.

From the laboratory book, dated November 28 and 29, 1825, he at this time worked on electric and electro-magnetic induction.

'Two copper wires were tied close together, a thickness of paper only intervening, for a length of five feet: one of them was made the connecting wire of a battery of forty pairs of plates, four inches square, in rather weak action; and the ends of the other were connected with a galvanometer. No effects, however, upon its needle could be observed; consequently no visible proofs of induction by the wire, through which the current was passing, upon its neighbour could this way be perceived.

'In reference to certain views with respect to the axis of action; the connecting wire of the battery passed through the centre of a helix, but this gave no results. Again, a helix being in connection with the poles of the pile, a straight wire occupying its axis was connected with the galvanometer, but no apparent effects (were observed). The galvanometer was not a very delicate one.'

'If magnetic action be simply electrical action, as M. Ampère considers it, then magnetic induction must be electrical induction, and M. Arago's experiments must depend upon induced electrical action. Hence electrical poles or surfaces ought to produce similar effects; for though the electricity will not be in such quantity, and not in motion, still it has sufficient attractive and repulsive powers. And it appears to me that the mere difference of motion or rest, as respects the electricity in the inducing body, will not explain such retention of the induced state in one case (magnetism), and such resignation of it in the other (the experiment proposed), as to account for dragging attraction in the former, and not in the latter state of things.'

'De Luc's column well warmed and suspended by a silk thread, five feet long, over a plate of copper—revolution of the plate caused no revolution of the piles.'

'A Leyden jar was fitted with a wire and ball, then suspended upside down, so that when charged its knob was positive, the knob of the wire negative; it was then brought over opposite extremities of the diameter of the wheeling copper plate. No difference, however, could be perceived in the action of the plate when in motion upon the jar and wire, whether the latter were charged or uncharged.'

He prepared some experiments on the existence of vapour at low temperatures, and he sent all the apparatus necessary with Captain Franklin. Some years afterwards, he wrote at the end of his notes, 'Never got any account of the results.'

III.

His lectures, appointments, and titles, in 1824–25 were these:—

Professor Brande at this time gave a course of lectures, in the morning, on chemistry and physics. Faraday took a part of this course of laboratory lectures in 1824.

The President and Council of the Royal Society appointed a committee for the improvement of glass for optical purposes. It consisted of Fellows of the Royal Society and members of the Board of Longitude. Faraday was put on this committee.

Early in January 1824 he was elected a Fellow of the Royal Society; also of the Geological Society, and honorary member of the Cambrian Society of Swansea. In 1825 he was elected a member of the Royal Institution and a corresponding member of the Society of Medical Chemists, Paris.

From the manager's minutes for February 7, 1825, it appears that Sir H. Davy, 'having stated that he considered the talents and services of Mr. Faraday, assistant in the laboratory, entitled to some mark of approbation from the managers, and these sentiments having met the cordial concurrence of the board: Resolved that Mr. Faraday be appointed Director of the Laboratory under the superintendence of the Professor of Chemistry.'

His first act showed his energy and desire to promote the welfare of the members. He invited the members of the Institution to come to evening meetings in the laboratory. 'Three or four meetings took place this year.' At one of these Faraday gave the members an account of the electro-magnetic motions which he had discovered four years previously. From these evenings in the laboratory the present Friday evening discourses in the theatre of the Institution had their origin.

In May, a sub-committee, consisting of Mr. Herschel,

Mr. Dollond, and Mr. Faraday, was appointed to have the direct superintendence and performance of experiments on the manufacture of optical glass. 'It was my business to investigate particularly the chemical part of the inquiry: Mr. Dollond was to work and try the glass, and ascertain practically its good or bad qualities; whilst Mr. Herschel was to examine its physical properties, reason respecting their influence and utility, and make his competent mind bear upon every part of the inquiry. In March 1829 the committee was reduced to two by the retirement of Mr. Herschel, who about that period went to the Continent.'

In July he left London by steamboat for Scotland. After visiting the damask works at Edinburgh, he saw the glass works at Leith. He minutely describes the geology of Salisbury Craig, Arthur's Seat, and Craigsleith quarries; and then he went to Rubislaw (Bleaching Liquor Works), Aberdeen. Here he made many experiments for the proprietors, with whom he stayed.

IV.

July 31, 1824.—He wrote from London to Mrs. Faraday, at Niton, in the Isle of Wight.

He had left her two days before; slept at Freshwater Gate, crossed next day from Yarmouth to Lymington, got to Southampton in the evening, and reached London next morning by eight o'clock.

FARADAY TO MRS. FARADAY.

'Saturday, July 31, 1824.

'My dear Sarah,—The preparations I make seem to promise you what you desired—namely, a very long descriptive letter; and if I can keep my eyes open (and whilst writing to you there can be no reason to doubt that), you shall have your desire. So much occurred in the various little incidents after I left you that I wished to tell you, that I think I had better go on in regular order from that time till the present moment.

. . . . . .

'I feel rather tired and stiff myself, and perhaps that makes my letter so too; but my dear girl is, I know, a girl of consideration, and will not insist upon having two or three pages of affection after so much narrative. Indeed, I see no use in measuring it out at all. I am yours, my heart and thoughts are yours, and it would be a mere formality to write it down so; and capable of adding nothing to the truth, but that I have as much pleasure in saying it as you have in hearing it said, and that it is not with us at least a measure or token of affection merely, but the spontaneous result of it. I have not yet been to see my mother, but I am going immediately.

. . . . . .

'I found certain French and German journals here, and, on inquiry at Murray's, found that Dr. Ure had given up his department of the Journal (i.e. I fancy it has been taken from him), and the journals were sent to me to assist the Miscellanea. The Miscellanea swim as long as most of the departments of that journal.

. . . . . .

'Adieu, my dear girl, for the present: write to me soon; give my love to father and mother, and remember me to Mr. L. I shall write to you again about the end of next week.

'From your sincerely affectionate husband,

'M. FARADAY.'

His description and remarks upon Brighton, in a letter to his sister, show something of his character and observation.

FARADAY TO HIS YOUNGEST SISTER.

'Niton: August 25, 1824.

'My dear Margaret,—. . . At Brighton we dined with sister S., and found all well; we then rambled out geologising, &c.

'I do not at all admire Brighton, i.e. its character as a fashionable or interesting place. It is a very convenient place for distance, lodging, accommodation, food, &c.—but these are not what I refer to; I mean as to its beauties, natural or artificial, or as to its importance in advancing great interests, as civilisation or improvement.

'Considered in this way, Brighton is to me very commonplace and poor: there are no natural beauties there to distinguish it from a thousand other places; there are no high interests concerned to raise it above the poor distinction of being a place resorted to by company because other company was there before them. As to the Pavilion, there is scarcely a single cottage in or about this poor village of Crab Niton that does not both in beauty and use surpass it. The Pavilion has no beauty for the painter; and what is intended for beauty, of which there is a great deal, has no use.

'The Steine is a good street, and many of the squares and places are good, also many of the old houses; and could one but see a sufficient cause why they had come together—i.e. the presence of any beautiful or useful feature—the town, with the exception of one or two things, would be a very good one. It has, however, one thing perfectly beautiful every way in the chain-pier—an admirable specimen of ingenuity and art, and which, destined to useful purposes, not only pleases the eye but satisfies the mind.

. . . . . .

'Your ever affectionate brother,

'M. FARADAY.'

From 1826 to 1829, the knowledge that remains of (1) the scientific education, (2) the reputation, and (3) the character of Faraday comes from the same sources as in previous years.

I.

In 1826 he had two papers in the 'Philosophical Transactions:' one was upon the mutual action of sulphuric acid and naphthaline, and the discovery of sulpho-naphthalic acid; and the other was on the existence of a limit to vaporisation. These were the chief subjects of his laboratory work during this and the previous year.

In the 'Quarterly Journal' he had ten papers. Of these the chief were on pure caoutchouc, and the substances by which it is accompanied; on sulpho-naphthalic acid; on bisulphuret of copper; and on the fluidity of sulphur at common temperatures.

He began the Friday evening meetings of the members of the Royal Institution. Out of the seventeen discourses this year Faraday gave six. These were,—on pure caoutchouc; on Brunel's condensed gas-engine; on lithography; on sulpho-vinic and sulpho-naphthalic acid; on Drummond's light; on Brunel's tunnel at Rotherhithe.

At the beginning of the third evening he gave

from the following notes his idea of the nature of these lectures: 'Evening opportunities—interesting, amusing, instruct also:—scientific research—abstract reasoning, but in a popular way—dignity;—facilitate our object of attracting the world, and making ourselves with science attractive to it.' In another notebook he made a list of no less than fifty-four subjects for single lectures, the object being 'to illustrate popular subjects, to connect parts of science and facts generally separated and sometimes neglected in scientific arrangement;' and in the same book he made a list of lecture experiments and illustrations. This he continued down to 1850.

In 1827 he published his 'Chemical Manipulation,' in one volume, 8vo; a second edition appeared in 1830, and a third in 1842. He refused to bring out a later edition, although it would have continued to be profitable.

The object of the volume is to facilitate to the young chemist the acquirement of manipulation, and, by consequence, his progress in the science itself. It does not attempt to inculcate the principles of the science, but the practice; neither does it claim to teach a habit of reasoning, but has solely in view the art of experimenting. In the pursuit of this object it is intended to describe:—

The conveniences and requisites of a laboratory.
General chemical apparatus and its uses.
The methods of performing chemical operations.
The facilities acquired by practice.
The causes which make experiments fail or succeed.

He had six papers in the 'Quarterly Journal of Science:'—On the probable decomposition of certain gaseous compounds of carbon and hydrogen during sudden expansion. On a peculiar perspective appearance of aerial light and shade. Experiments on the nature of Labarraque's disinfecting liquid. On the fluidity of sulphur and phosphorus at common temperatures. On transferrence of heat by change of capacity in gas. On the confinement of dry gases over mercury. He begins this last paper thus:—'The results of an experiment made June 26th, 1823, by myself, and quoted as such, having been deemed of sufficient interest to be doubted, I have been induced to repeat it.' And he then goes on to give three of the earliest experiments that were ever made on the diffusion of gases.

There is a sad interest attached to these experiments. The doubter was Sir Humphry Davy. The last experiment Davy made in the laboratory of the Royal Institution was probably on this subject; for the last entry in the laboratory book in his handwriting is dated Feb. 5, 1826. 'Experiment of July 1825 examined.—The hydrogen confined by mercury, whether in a bottle with a stopper, or merely confined by dry mercury, contained no common air.

'In that confined by water and mercury there was a slight diminution, such as might be expected from the air contained in the water, but not appreciable.'

Faraday's experiments were begun, he says, on June 28, 1825, and the examination of the gases was made September 15, 1826. He proved that there was no mistake in his original observation made in 1823–24.

In the paper on sulphur he says, 'I published some time ago (the year previous) a short account of an instance of the existence of fluid sulphur at common temperatures; and though I thought the fact curious, I did not esteem it of such importance as to put more than my initials to the account. I have just learned, through the "Bulletin Universel" for September (p. 78), that Signor Bellàni had observed the same fact in 1813, and published it in the "Giornale di Fisica." M. Bellani complains of the manner in which facts and theories which have been published by him are afterwards given by others as new discoveries; and though I find myself classed with Gay-Lussac, Sir H. Davy, Daniell, and Bostock, in having thus erred, I shall not rest satisfied without making restitution, for M. Bellani in this instance certainly deserves it at my hand.'

He gave his first course of six lectures in the theatre of the Institution in April, on chemical philosophy.

His subjects were, the atmosphere, general view of the relation of air, gases, and vapour; chemical affinity, definite proportions, flame, voltaic pile, magnetism principally as evolved by electricity.

In his second lecture he says:—'We may now boldly affirm that no philosophical reason exists for making a distinction between gases and vapours. Gases are shown not to be permanently elastic, but to be condensible like acknowledged vapours.'

He ends his third lecture thus:—'Wonderful activity of matter in nature from few chemical elements. Our present state a state of quiescence almost. Conceive the effect of putting the elements of the globe together in an uncombined state, or even of slightly altering the proportion of the affinities, which in their present state are so admirably and beautifully arranged as to give energy to the volcano, to be subservient to the expansion of the tenderest bud or flower, and to minister equally and essentially to the development and the existence of the most perfect and the most transitory of animated beings.'

He gave twelve lectures, from February to May, at the London Institution. The subject of his course was on the philosophy and practice of chemical manipulation.

He began his first lecture thus, on February 13:—

'The object for which we are assembled will have been announced to you by the notices and the cards that have been issued from the authorities of this Institution. It is the development, in a course of lectures, of the *principles and practice of chemical manipulation.* The subject is new to the lecture-room, and almost to the library; nor has it ever been considered in that full and explicit manner it deserves. With the exception of some brief and general directions, each person has been left to discover and work out the means comprised in it for himself; and that which is essential to the progress of everyone in chemical science has been taught only, in the very depths of the laboratory, to a highly privileged few, more indeed in the manner of an alchemical secret than of useful and experimental knowledge which should be open to all.

'You will not, therefore, be surprised at the use of terms which may now and then sound new to your ears, though I trust always clear to your understandings. The word manipulation, for instance, though not usual in ordinary language, is so peculiarly expressive of the great object of these lectures, that I could not hesitate a moment to use it. It implies working, or, more precisely, the use of the hands, being derived from the word *manus*, which signifies a hand;

and by the expression "chemical manipulation" I wish you to understand that practice and habit of using the hands expertly in chemical investigation by which the philosopher may successfully acquire experimental truth. It is curious that the only recent dictionary which notices the word has given a very partial and limited meaning in place of the general one; and if upon this occasion you feel inclined to refer to such authorities, I must beg to direct your attention to the "Dictionnaire de Trevoux," where its sense and meaning are correctly expressed.

'On entering upon this subject, I for my own sake beg to give a direction to your expectations, while I disclaim any endeavour to furnish you in twelve lectures with that kind and degree of manipulatory knowledge which is necessary to the philosopher or even to the student. I should cheat you were I, for the sake of attracting your attention, to lead you to suppose this could be done. Instruction by lectures being confined to the few and limited occasions upon which the lecturer and his audience meet, is deficient in those essential elements of expert manipulation—time and practice.

'Indeed, if it were possible, I should presume to offer such information only to a bench of students; for I think I shall better fill the office with which I am here honoured by illustrating and explaining the means which others take to wrest truths from nature, than by insufficiently prompting you to imitate their exertions. I intend to take advantage of the multitude of beautiful facts that have been discovered to illustrate the manner of their development. I desire to make you acquainted with the means by which chemical science is primarily advanced and exalted, chemical arts extended and improved; and by laying open to you the paths through which others are running their career, enable you to watch their advances, judge their assertions, estimate their difficulties, and award their praise.'

On December 29 he gave, at the Royal Institution, the first of a course of six lectures on chemistry, adapted to a juvenile audience. His notes began thus: 'Substances and affinity; brief remarks upon the objects of the course. Touch principally upon tangible chemistry, and then only on those parts which, being constantly before us in one form or another, ought to be well understood in the first place, if only as being part of general knowledge. Desire to convey clear notions of some of the most important and familiar chemical objects around us. Intended simplicity of the lectures.' There are eighty-six experiments put down to illustrate this first lecture. The last words of the last lecture are, 'Now take leave, hoping you will remember a good deal of what I have told you and shown you respecting the atmosphere, water, combustible bodies, the acids and metals, those very important elements of chemical knowledge.'

Afterwards he added in his note-book, 'These six juvenile lectures were just what they ought to have been, both in matter and manner, but it would not answer to give an extended course in the same spirit.'

In another note he says: 'This year the President and Council of the Royal Society applied to the President and Managers of the Royal Institution for leave to erect on their premises an experimental room with a furnace, for the purpose of continuing the investigation on the manufacture of optical glass. They were guided in this by the desire which the Royal Institution has always evinced to assist in the advancement of science, and the readiness with which the application was granted showed that no mistaken notion had been formed in this respect. As a member of both bodies, I felt much anxiety that the investigation should be successful. A room and furnaces were built at the Royal Institution in September 1827, and an assistant was engaged, Sergeant Anderson, of the Royal Artillery. He came on December 3.

There were nineteen Friday evening meetings at the Royal Institution. Faraday gave three discourses.

The first was on magnetic phenomena developed by metals in motion. At the end of the notes of this lecture Faraday again says what he wished the Friday evenings to be—'permitted to refer to them—as being actively engaged in their first institution, and as secretary of the Committee — their nature — agreeable — easy—meeting—where members have the privilege of bringing friends, and where all may feel at ease—desirable to have all things of interest placed there—large or small—opportunities of library or lecture room—nature of the lecture room affair—relieved from all formalities except those essential to secure the attention and freedom of all—long or short—good matter—the kind —after which adjourn to tea and talk—may well hope that now the feeling such, that literary subjects shall be intermingled with those of science and art.'

In 1828 he published two papers in the 'Quarterly Journal:' on the relation of water to hot polished surfaces, and on anhydrous crystals of sulphate of soda.

He gave eight lectures, 'On the operations of the laboratory,' after Easter this year. He says, 'They were not to my mind. There does not appear to be that opportunity of fixing the attention of the audience, by a single, clear, consistent, and connected chain of reasoning, which occurs when a principal or one particular application is made. The lectures appeared to me to be broken, or, at least, the facts brought forward were not used as proofs of their most striking or important effects, but as proofs of some subordinate effect common to all. I do not think the operations of the laboratory can be rendered useful or popular at the same time in lectures; or, at least, I think I have not the way, and can do better with other subjects, as some general points of chemical philosophy.'

He gave five of the Friday evening lectures: illustrations of the new phenomena produced by a current of air or vapour recently observed by M. Clement; two on the reciprocation of sound—the matter belonged to Mr. Wheatstone, but was delivered by Faraday, as was also a discourse on the nature of musical sound; the last evening was on the recent and present state of the Thames Tunnel.

In 1829 he gave the Bakerian lecture at the Royal Society, on the manufacture of glass for optical purposes.

This most laborious investigation did not end in the desired improvement in telescopes; but the glass then manufactured, as will be seen hereafter, became of the utmost importance in Faraday's diamagnetic and magneto-optical researches, and the work led to the permanent engagement, in 1832, of Mr. Anderson as his assistant.

This lecture is printed in the 'Philosophical Transactions' for 1830. At the beginning Faraday says: 'I cannot resist the occasion which is thus offered to me of mentioning the name of Mr. Anderson, who came to me as an assistant in the glass experiments, and has remained ever since in the laboratory of the Royal Institution. He has assisted me in all the researches into which I have entered since that time; and to his care, steadiness, exactitude, and faithfulness in the performance of all that has been committed to his charge, I am much indebted.'

He gives the following introduction :—

'When the philosopher desires to apply glass in the construction of perfect instruments, and especially the achromatic telescope, its manufacture is found liable to imperfections so important and so difficult to avoid, that science is frequently stopped in her progress by them—a fact fully proved by the circumstance that Mr. Dollond, one of our first opticians, has not been able to obtain a disc of flint glass, 4½ inches in diameter, fit for a telescope, within the last five years; or a similar disc, of 5 inches, within the last ten years.

'This led to the appointment by Sir H. Davy of the Royal Society Committee, and the Government removed the excise restrictions, and undertook to bear all the expenses as long as the investigation offered a reasonable hope of success.

'The experiments were begun at the Falcon Glass Works, three miles from the Royal Institution, and continued there in 1825, 1826, and to September 1827, when a room was built at the Institution. At first the inquiry was pursued principally as related to flint and crown glass; but in September 1828 it was directed exclusively to the preparation and perfection of peculiar heavy and fusible glasses, from which time continued progress has been made.'

The paper then proceeds with an exact description

of this heavy optical glass: 'Its great use being to give efficient instructions to the few who may desire to manufacture optical glass.'

In 1830 the experiments on glass-making were stopped.

In 1831, the Committee for the Improvement of Glass for Optical Purposes reported to the Royal Society Council 'that the telescope made with Mr. Faraday's glass has been examined by Captain Kater and Mr. Pond. It bears as great a power as can reasonably be expected, and is very achromatic. The Committee therefore recommend that Mr. Faraday be requested to make a perfect piece of glass of the largest size that his present apparatus will admit, and also to teach some person to manufacture the glass for general sale.'

In answer to this, Faraday sent the following letter to Dr. Roget, Sec. R. S.

'Royal Institution: July 4, 1831.

'Dear Sir,—I send you herewith four large and two small manuscript volumes relating to optical glass, and comprising the journal book and sub-committee book, since the period that experimental investigations commenced at the Royal Institution.

'With reference to the request which the Council of the Royal Society have done me the honour of making —namely, that I should continue the investigation—I should, under circumstances of perfect freedom, assent to it at once; but obliged as I have been to devote the whole of my spare time to the experiments already described, and consequently to resign the pursuit of such philosophical inquiries as suggested themselves to my own mind, I would wish, under present circumstances, to lay the glass aside for a while, that I may enjoy the pleasure of working out my own thoughts on other subjects.

'If at a future time the investigation should be renewed, I must beg it to be clearly understood I cannot promise full success should I resume it: all that industry and my abilities can effect shall be done; but to perfect a manufacture, not being a manufacturer, is what I am not bold enough to promise.

'I am, &c.,

'M. FARADAY.'

In 1845 he added this note :—

'I consider our results as negative, except as regards any good that may have resulted from my heavy glass in the hands of Amici (who applied it to microscopes), and in my late experiments on light.'

In May 1829, at the Institution, he gave six lectures on various points of chemical philosophy. His subjects were, water, hydrocarbons, artificial heat, artificial light, safety lamp, common salt.

He ended his lecture on the safety lamp with the words he had used in the lecture on this subject at the City Philosophical Society: 'Such is the philosophical history of this most important discovery. I shall not refer to supposed claims of others to the same invention, more than to say that I was a witness in our laboratory to the gradual and beautiful development of the train of thought and experiments which produced it. The honour is Sir H. Davy's, and I do not think that this beautiful gem in the rich crown of fame which belongs to him will ever be again sullied by the unworthy breath of suspicion.'

He gave a Friday evening discourse on Mr. Robert Brown's discovery of active molecules in bodies, either organic or inorganic. He ends the notes of this lecture thus :—

'*Lastly, the relation of these appearances to known or unknown causes.* Analogy to other moving particles. Camphor. Supposed facility of explanation, *not camphor motion.* Takes place within pollen. Under water, enclosed by mica or oil—not *crystalline* particles—not attraction or repulsion. Does not consist in receding and approaching—*not evaporation* answered as before —not currents too minute—oscillation—consider current in a drop—when currents, motion very different—not electricity of ordinary kind, because do not come to rest, seen after hours—so *that the cause is at present undetermined.*

'Mr. Brown, supposed to be careless and bold, is used to microscopical investigations—has not yet been corrected—assisted by Dr. Wollaston—so that carelessness can hardly be charged. Then, what does Mr. Brown say? simply that he cannot account for the motions.

'Many think Mr. Brown has said things which he has not—but that is because the subject connects itself so readily with general molecular philosophy that all *think* he must have meant this or that—as to *molecules*, by no means understand ultimate atoms—as to size, says that solid matter has a tendency to divide into particles about that size—pulverisation and precipitation—if smaller, which may be, are *very difficult to see*—does not say that all particles are alike in their nature, but simply that organic and inorganic particles having motion, motion cannot be considered as distinctive of vitality—connection with atomic or molecular philosophy.'

He gave five other Friday discourses: on Brard's test of the action of weather on building stones; on Wheatstone's further investigations on the resonances of reciprocal vibrations of volumes of air; on Brunel's block machinery at Portsmouth; on the phonical or nodal figures at vibrating surfaces; on the manufacture of glass for optical purposes.

At Christmas he gave a course of juvenile lectures on electricity. His notes begin thus:—

'An extraordinary power that I have to explain; not fear boldly entering into its consideration, because I think it ought to be understood by children—not minutely, but so as to think reasonably about it, and such effects as children can produce, or observe to take place in nature—simple instances of its power.'

He wrote down eighty experiments for this first lec ture.

II.

The increase of his reputation during 1826, 1827, 1828, and 1829 is seen in his titles and in the appointments offered to him.

In 1826 he was made an honorary member of the Westminster Medical Society, and the managers of the Royal Institution 'relieved him from his duty as chemical assistant at the lectures because of his occupation in research.'

In 1827 he was made correspondent of the Société Philomathique, Paris.

In 1827 he was offered the Professorship of Chemistry in the new University of London, which then consisted only of University College. The letter to Dr. Lardner in which he declines the appointment shows his great attachment to the Royal Institution.

FARADAY TO DR. LARDNER.

'Royal Institution: October 6, 1827.

'My dear Sir,—My absence from town for a few days has prevented your letter from receiving an answer so soon as it ought to have done; and to compensate for the delay I should have called upon you yesterday evening, but that I prefer writing in the present case, that my reasons for the conclusion at which I have arrived may be clearly stated and understood.

'You will remember, from the conversation which we have had together, that I think it a matter of duty and gratitude on my part to do what I can for the good of the Royal Institution in the present attempt to establish it firmly. The Institution has been a source of knowledge and pleasure to me for the last fourteen years, and though it does not pay me in salary for what I *now* strive to do for it, yet I possess the kind feelings and good-will of its authorities and members, and all the privileges it can grant or I require; and, moreover, I remember the protection it has afforded me during the past years of my scientific life. These circumstances, with the thorough conviction that it is a useful and valuable establishment, and the strong hopes that exertions will be followed with success, have decided me in giving at least two years more to it, in the belief that after that time it will proceed well, into whatever hands it may pass. It was in reference to this latter opinion, and fully conscious of the great opportunity afforded by the London University of establishing a valuable school of chemistry and a good name, that I have said to you and Mr. Millington, that if things altogether had been two years advanced, or that the University had to be founded two years hence, I should probably have eagerly accepted the opportunity. As it is, however, I cannot look forward two years and settle what shall happen then. Upon general principles only I should decline making an engagement so long in advance, not knowing what might in the meantime occur; and as it is, the necessity of remaining free is still more strongly urged upon me. Two years may bring the Royal Institution into such a state as to make me still more anxious to give a third to it. It may just want the last and most vigorous exertions of all its friends to confirm its prosperity, and I should be sorry not to lend my assistance with that of others to the work. I have already (and to a great extent for the sake of the Institution) pledged myself to a very laborious and expensive series of experiments on glass, which will probably require that time, if not more, for their completion; and other views are faintly opening before us. Thus you will see, that I cannot with propriety accede to your kind suggestion.

'I cannot close this letter without adverting to the honour which has been done me by my friends, and I may add by the Council of the University, in their offering me the chemical chair in so handsome and unlimited a manner; and, if it can be done with propriety, I wish you to express my strong sentiments on this point to those who have thought of me in this matter. It is not the compliment and public distinction (for the matter is a private one altogether), but the high praise and approbation which such an unlimited mark of their confidence conveys, and which, coming to me from such a body of men, is more valuable than an infinity of ordinary public notice. If you can express for me my thankfulness for such kind approbation, and the regret which I feel for being obliged by circumstances to make so poor a return for their notice, I shall be much obliged to you.

'You will remember that I have never considered this affair except upon general views, for I felt that unless these sanctioned my acceptance of the Professorship, it would be useless to inquire after such particulars as salary, privileges, &c. I make this remark now, that you may not suppose these have been considered and approved of, supposing other things had been favourable. I have never inquired into them, but from general conversation have no doubt they would have proved highly satisfactory.

'I am, my dear Sir, yours very truly,

'M. FARADAY.'

In 1828 he was made a Fellow of the Society of Natural Science of Heidelberg.

He was invited to attend the meetings of the Board of Managers of the Institution.

And he received his first medal, which was founded by Mr. Fuller, a member of the Royal Institution.

In 1829 he was made a member of the Scientific Advising Committee of the Admiralty; patron of the Library of the Institution; honorary member of the Society of Arts, Scotland.

In 1829 he was asked to become lecturer at the Royal Academy, Woolwich. His letter to Colonel

Drummond, in which he accepts the appointment, is also very characteristic.

FARADAY TO COLONEL DRUMMOND.

'Royal Institution: June 29, 1829.

'Sir,—In reply to your letter of the 26th, and as a result of our conversation on Saturday, I beg to state that I should be happy to undertake the duty of lecturing on chemistry to the gentlemen cadets of Woolwich, provided that the time I should have to take for that purpose from professional business at home were remunerated by the salary.

'But on this point I hardly know what to say in answer to your inquiry, because of my ignorance of the conveniences and assistance I should find at Woolwich. For the lectures which I deliver in this Institution, where I have the advantage of being upon the spot, of possessing a perfect laboratory with an assistant in constant occupation, and of having the command of an instrument maker and his men, I receive, independent of my salary as an officer of the establishment, 8*l*. 15*s*. per lecture. The only lectures I have given out of this house were a course at the London Institution, for which, with the same conveniences as to laboratory and assistance, I was paid at the same rate. Since then I have constantly declined lecturing out of the Royal Institution because of my engagements.

I explained to you on Saturday the difficulty of compressing the subject of chemistry into a course of twenty lectures only, and yet to make it clear, complete, and practically useful; and without I thought I could do this, I should not be inclined to undertake the

1826–29. Æt. 34–38.

charge you propose to me. Now twenty lectures, at the terms I have in this house, amount to 175*l.* per annum, and therefore I should not be inclined to accept any offer under that; the more especially as, if I found that the times and hours of the students allowed it, I should probably extend the course by a lecture or two, or more, that I might do the subject greater justice.

'Notwithstanding what I have said, I still feel the difficulty of estimating labour, the extent of which I am ignorant of.

'Were the lectures of that class which do not require to be accompanied by experiment, or were the necessary experiments and illustrations of such a nature that (as in mechanics) the preparations, once made, are complete and ready when wanted for future courses, I should not feel the difficulty. But in many parts of chemistry, and especially in the chemistry of the gases, the substances under consideration cannot be preserved from one course to another, but have to be formed at the time; and hence, if the illustrations are to be clear and numerous, a degree of preparatory labour, which has to be repeated on every occasion.

'For these reasons I wish you would originate the terms rather than I. If you could make the offer of 200*l.* a year, I would undertake them; and then, supposing I found more work than I expected, I should not have to blame myself for stating an undervalue for my own exertions. I have no thought that the sum would overpay, because, from my experience for some years in a chemical school founded in the laboratory of the Royal Institution, I have no doubt that the proportion of instruction to the students would expand rather than contract.

'Allow me, before I close this letter, to thank you and the other gentlemen who may be concerned in this appointment, for the good opinion which has induced you to propose it to me. I consider the offer as a high honour, and beg you to feel assured of my sense of it. I should have been glad to have accepted or declined it, independent of pecuniary motives; but my time is my only estate, and that which would be occupied in the duty of the situation must be taken from what otherwise would be given to professional business.

'I am, Sir, your most obedient servant,

'M. FARADAY.'

III.

The correspondence that remains of these four years consists only of two letters.

They were both written on the same day—one to his brother-in-law, who was at that time much depressed, and the other to his friend Magrath.

The contrast of the tone of these letters is striking, but both show the kind feeling that was in him.

FARADAY TO HIS BROTHER-IN-LAW, E. BARNARD.

'Niton: July 23, 1826.

'My dear Edward,—I intended to have written you a letter immediately upon the receipt of yours, but delayed it, and perhaps shall not now say what occurred to me then. Why do you write so dully? Your cogitations, your poetry, and everything about your letter, except the thirty pounds, has a melancholy feel. Perhaps things you had scarcely anticipated are gathering about you, and may a little influence your spirits; and I shall think it is so for the present, and trust it is of but little importance, for I can hardly imagine it possible that you are taken unawares in the general picture of life which you have represented to yourself: your natural reflection and good sense would teach you that life must be chequered, long before you would have occasion to experience it. However, I shall hope this will find you in good spirits, and laughing at such thoughts as those in which you were immersed when you wrote me. I have been watching the clouds on these hills for many evenings back: they gather when I do not expect them; they dissolve when, to the best of my judgment, they ought to remain; they throw down rain to my mere inconvenience, but doing good to all around; and they break up and present me with delightful and refreshing views when I expect only a dull walk. However strong and certain the appearances are to me, if I venture an internal judgment, I am always wrong in something; and the only conclusion that I can come to is, that the end is as beneficial as the means of its attainment are beautiful. So it is in life; and though I pretend not to have been much involved in the fogs, mists, and clouds of misfortune, yet I have seen enough to know that many things usually designated as troubles are merely so from our own particular view of them, or else ultimately resolve themselves into blessings. Do not imagine that I cannot feel for the distresses of others, or that I am entirely ignorant of those which seem to threaten friends for whom both you and I are much concerned. I do feel for those who are oppressed either by real or imaginary evils, and I know the one to be as heavy as the other. But I think I derive a certain degree of steadiness and placidity amongst such feelings by a point of mental conviction, for which I take no credit as a piece of knowledge or philosophy, and which has often been blamed as mere apathy. Whether apathy or not, it leaves the mind ready and willing to do all that can be useful, whilst it relieves it a little from the distress dependent upon viewing things in their worst state. The point is this: in all kinds of knowledge I perceive that my views are insufficient, and my judgment imperfect. In experiments I come to conclusions which, if partly right, are sure to be in part wrong; if I correct by other experiments, I advance a step, my old error is in part diminished, but is always left with a tinge of humanity, evidenced by its imperfection. The same happens in judging of the motives of others; though in favourable cases I may see a good deal, I never see the whole. In affairs of life 'tis the same thing; my views of a thing at a distance and close at hand never correspond, and the way out of a trouble which I desire is never that which really opens before me. Now, when in all these, and in all kinds of knowledge and experience, the course is still the same, ever imperfect to us, but terminating in good, and when all events are evidently at the disposal of a Power which is conferring benefits continually upon us, which, though given by means and in ways we do not comprehend, may always well claim our acknowledgment at last, may we not be induced to suspend our dull spirits and thoughts when things look cloudy, and, providing as well as we can against the shower, actually *cheer our spirits* by thoughts of the good things it will bring with it? and will not the experience of our past lives convince us that in doing this we are far more likely to be right than wrong.

'Your third page I can hardly understand. You quote Shakespeare; the quotation may be answered a thousand times over from a book just as full of poetry, which you may find on your shelf. The uses of the world can never be unprofitable to a reflecting mind, even without the book I refer to; and I am sure can only appear so to you for a few hours together. But enough of this; only, when I get home again, I must have a talk with you.

. . . .

'Believe me, my dear Edward, your affectionate brother,

'M. FARADAY.'

FARADAY TO MAGRATH.

'Niton: July 23, 1826.

'I am amused and a little offended at ——'s hypocrisy. He knows well enough that to the world an hour's existence of our Institution is worth a year's of the ——, and that though it were destroyed, still the remembrance of it would live for years to come, in places where the one he lives at has never been heard of. Unless he comes with perfect good will and feeling in every part of the way, I do not think I at least shall meet him; for that nonsense of his, though it may amuse once or twice or thrice, becomes ridiculous if it is to be thrown into every affair of life, both common and serious, and would probably be in our way. I think it would not be a bad joke to touch him up behind, and say one can't imagine how it is that he is only assistant librarian at such an unknown institution as the ——, and that one can't help but imagine there must be some cause or other, or he would be aiming at a higher character in the house, or would endeavour to get into a more public place, &c. I think I could make the man wince if I were inclined, and yet all in mere chat over a cup of tea. But this is all nonsense, which, however, he brings to mind by the corresponding nonsense of his own affectation.

'Now Hennel is a plain, common-sense man, without any particular varnish over his conduct and manners, and when he speaks one knows what he means. I feel much, therefore, for his disappointment, and think it altogether an unwise thing in some to be so neglectful of his desires and feelings as in this case they have been. Why should not we philosophers tempt recruits by every honourable means? And when Hennel had so worthily earned the reward of pleasurable feelings, why should they not be gratified when it might have been done with so little trouble? It annoys me as much, I think, as it will Hennel himself, for I felt a great anxiety to see a copy of *his first paper* to the Royal Society. . . . .

'I am, dear Magrath, very truly yours,

'M. FARADAY.'

In 1830 the higher scientific education of Faraday was nearly ended. The records of this year show but little of his work, of his reputation, or of his character.

I.

With regard to his work, his Bakerian lecture on glass was published in the 'Philosophical Transactions.'

He had no lectures to give after Easter at the Royal Institution

His Friday evening discourses were on Aldini's proposed method of preserving men exposed to flame; on the transmission of musical sounds through solid conductors and their subsequent reciprocation; on the flowing of sand under pressure; on the measurement of a base in Ireland for the geodetical survey; on the application of a new principle in the construction of musical instruments; on the laws of coexisting vibrations in strings and rods, illustrated by the kaleidophone.

II.

The reputation which he was gaining abroad is seen in two letters which he received from members of the French Academy, M. Hachette and M. Ampère. The former gives Mr. Faraday an account of the Revolution of 1830, and his opinion of the influence of science, in words which must have sounded very exaggerated to one who throughout his life took only the slightest interest in politics.

M. HACHETTE TO FARADAY.

'Paris: 22 août 1830.

'Monsieur,—Vous avez probablement reçu un petit mémoire sur des expériences hydrauliques que je vous ai envoyé le 17 juillet passé, en même temps que je reçus votre mémoire sur le verre; un autre mémoire de M. Davies Gilbert s'y trouvait joint. J'ai traduit ce dernier mémoire, et la traduction sera publiée dans le bulletin de la Société d'Encouragement, cahier de juillet. J'ai ajouté quelques notes à cette traduction, qui, j'espère, seront accueillies par M. Davies Gilbert.

'J'espérais pouvoir vous envoyer avec cette lettre

quelques exemplaires de mes notes précédées du mémoire, mais l'imprimeur du bulletin ne me les a pas encore renvoyées.

'Je profiterais de la première occasion pour me rappeler à votre souvenir et à celui de M. Davies Gilbert, qui a bien voulu me gratifier d'un exemplaire de son mémoire sur les machines à vapeur du Cornwall.

'Parmi les événements qui ont signalé les journées des 27, 28 et 29 juillet, vous aurez remarqué l'influence des sciences sur la population Parisienne. Des jeunes gens de l'âge moyen, 19 ans, formant l'École polytechnique, habitent un ancien collége placé aux extrémités de Paris: là ils étudient tranquillement les ouvrages de Lacroix, de Poissons, de Monge, etc.; l'analyse mathématique, la physique et la chimie enrichie de vos découvertes sont leur unique occupation. Un détachement armé se présente à eux et les invite à marcher avec lui pour la défense de la charte violée. Cette jeunesse humble, modeste, sans armes, revêtue de l'uniforme qui rappelle la défense de Paris en 1814, sort du collége, et à l'instant que chaque groupe arrive elle proclame un élève polytechnique son commandant. Elle serait invincible, puisque la science et l'honneur la précèdent; elle marche avec confiance, parce qu'elle a l'assentiment de tout ce que porte un cœur généreux. Les principes mathématiques (*principia mathematica*) et les principes de gouvernement se donnent donc la main: les deux premières nations du monde se rapprochent. Puisse la raison triompher de tous les préjugés qui s'opposent au perfectionnement des sociétés!

'En France, le savant, l'artiste, l'ouvrier sent toute la dignité de sa position sociale; chacun ajoute un peu de bien au bien qui existe; la plus petite découverte dans les sciences est un bienfait pour l'humanité ; les grandes découvertes sont pour notre siècle les actions héroïques. En vous exprimant mon opinion sur l'influence des sciences, j'éprouve un sentiment bien vif d'estime et de reconnaissance pour vous et vos compatriotes qui consacrez votre vie entière aux recherches scientifiques.

'J'ai l'honneur d'être bien affectueusement, Monsieur, votre dévoué serviteur,

'HACHETTE.'

M. AMPÈRE TO FARADAY.

'Paris: 13 octobre 1830.

'Monsieur et cher confrère,—Il y a bien longtemps que je devrais vous écrire : j'attendais d'avoir quelque chose de nouveau à vous offrir. Mais quoique je n'ai rien d'achevé dans ce moment je profite du voyage que va faire à Londres notre excellent ami Monsieur Unterwood pour vous rappeler les sentiments de la plus sincère amitié et de la reconnaissance que, comme dévoué aux sciences, j'éprouve pour l'auteur de tant de travaux qui ont agrandi et illustré sa carrière.

'La chimie et la physique vous doivent des résultats qui en font la gloire, et je vous dois personnellement au sujet des belles expériences sur les phénomènes de révolution et de rotation des aimants, qui sont venues justifier les recherches sur la cause que j'avais assignée aux phénomènes de ce genre. J'ai écrit un mémoire où j'ai développé avec beaucoup de détail tout ce qui est relatif ; je vous prie d'en accepter un exemplaire, que Monsieur Unterwood veut bien se charger de vous porter.

'C'est chez lui que je vous écris tout à la hâte, car il part demain pour Londres.

'Je vous prie d'agréer mes hommages et mes vœux pour que vous continuiez, dans l'intérêt des sciences, d'ajouter toujours à vos belles découvertes de nouvelles recherches dont les résultats leur fassent faire encore de nouveaux progrès.

'J'ai l'honneur d'être, Monsieur et cher confrère, votre très-humble et très-obéissant serviteur,

'A. AMPÈRE.'

III.

His brother-in-law, who then was much with him, and his niece, who formed one of the family at the Institution for nineteen years, have preserved some recollections of the period during which Faraday's higher scientific education went on.

Mr. George Barnard, the artist, says:—

'All the years I was with Harding I dined at the Royal Institution. After dinner we nearly always had our games just like boys—sometimes at ball, or with horse chestnuts instead of marbles—Faraday appearing to enjoy them as much as I did, and generally excelling us all. Sometimes we rode round the theatre on a velocipede, which was then a new thing.[1]

'At this time we had very pleasant conversaziones of artists, actors, and musicians at Hullmandel's, sometimes going up the river in his eight-oar cutter, cooking our own dinner, enjoying the singing of Garcia and his wife and daughter (afterwards Malibran)—indeed, of all the best Italian singers, and the society of most of

[1] Tradition remains that in the earliest part of a summer morning Faraday has been seen going up Hampstead Hill on his velocipede.

the Royal Academicians, such as Stanfield, Turner, Westall, Landseer, &c. . . . .

'My first and many following sketching trips were made with Faraday and his wife. Storms excited his admiration at all times, and he was never tired of looking into the heavens. He said to me once, "I wonder you artists don't study the light and colour in the sky more, and try more for effect." I think this quality in Turner's drawings made him admire them so much. He made Turner's acquaintance at Hullmandel's, and afterwards often had applications from him for chemical information about pigments. Faraday always impressed upon Turner and other artists the great necessity there was to experiment for themselves, putting washes and tints of all their pigments in the bright sunlight, covering up one half, and noticing the effect of light and gases on the other. . . . .

'Faraday did not fish at all during these country trips, but just rambled about geologising or botanising.'

Miss Reid, Mrs. Faraday's niece, writes thus:—

'About 1823, when my uncle was studying elocution under Smart, he took great trouble to teach me, a little girl of seven, to read with good emphasis, and I well remember how unweariedly he would go over and over one sentence, and make me repeat it with the upward and downward inflections, till he was satisfied; and then perhaps would follow a good romp, which pleased the little girl much better than elocution.

'After I went, in 1826, to stay at the Royal Institution, when my aunt was going out (as I was too little to be left alone), she would occasionally take me down to the laboratory, and leave me under my uncle's eye, whilst he was busy preparing his lectures. I had of course to sit as still as a mouse, with my needlework; but he would often stop and give me a kind word or a nod, or sometimes throw a bit of potassium into water to amuse me.

'In all my childish troubles, he was my never-failing comforter, and seldom too busy, if I stole into his room, to spare me a few minutes; and when perhaps I was naughty and rebellious, how gently and kindly he would win me round, telling me what he used to feel himself when he was young, advising me to submit to the reproof I was fighting against.

'I remember his saying that he found it a good and useful rule to listen to all corrections quietly, even if he did not see reason to agree with them.

'If I had a difficult lesson, a word or two from him would clear away all my trouble; and many a long wearisome sum in arithmetic became quite a delight when he undertook to explain it.

'I have a vivid recollection of a month spent at Walmer with my aunt and uncle. How I rejoiced to be allowed to go there with him! We went on the outside of the coach, in his favourite seat behind the driver. When we reached Shooter's Hill, he was full of fun about Falstaff and the men in buckram, and not a sight nor a sound of interest escaped his quick eye and ear. At Walmer we had a cottage in a field, and my uncle was delighted because a window looked directly into a blackbirds' nest built in a cherry-tree. He would go many times in a day to watch the parent birds feeding their young. I remember, too, how much he was interested in the young lambs, after they were sheared at our door, vainly trying to find their own mothers. The ewes, not knowing their shorn lambs, did not make the customary signal.

'In those days I was eager to see the sun rise, and my uncle desired me always to call him when I was awake. So, as soon as the glow brightened over Pegwell Bay, I stole down stairs and tapped at his door, and he would rise, and a great treat it was to watch the glorious sight with him. How delightful, too, to be his companion at sunset! Once I remember well how we watched the fading light from a hill clothed with wild flowers, and how, as twilight stole on, the sounds of bells from Upper Deal broke upon our ears, and how he watched until all was grey. At such times he would be well pleased if we could repeat a few lines descriptive of his feelings.

'He carried "Galpin's Botany" in his pocket, and used to make me examine any flower new to me as we rested in the fields. The first we got at Walmer was the *Echium vulgare*, and is always associated in my mind with his lesson. For when we met with it a second time he asked, "What is the name of that flower?" "Viper's Bugloss," said I. "No, no, I must have the Latin name," said he.

'One evening a thick white mist rose and completely hid everything before us. About ten o'clock my uncle called me into his room to see a spectre. He placed the candle behind us as we stood at the window, and there, opposite to us, appeared two gigantic shadowy beings who imitated every movement that we made.

'One of the first things to be done when he settled in the country was to set up a standing desk. It was made by putting the travelling boxes on a table. This was placed close to the window which was generally open, and the telescope was set up. There he wrote, but, however busy, nothing on sea or land escaped his eye. As he had gone to Walmer for rest and refreshment, I, the young one of the party, had to inveigle him away from his books whenever I could. Sometimes I was allowed to go to read with him, and my grandfather, who was staying with us, used to say, "What sort of reading lessons are those going on upstairs? I hear 'ha! ha!' more than any other sound."

'One day he went far out among the rocks, and brought home a great many wonderful things to show me; for in those days I had never seen nor even heard of hermit crabs and sea anemones. My uncle seemed to watch them with as much delight as I did; and how heartily he would laugh at some of the movements of the crabs! We went one night to look for glowworms. We searched every bank and likely place near, but not one did we see. On coming home to our cottage he espied a tiny speck of light on one of the doorposts. It came from a small centipede; but though it was put carefully under a glass, it never showed its light again.

'My uncle read aloud delightfully. Sometimes he gave us one of Shakespeare's plays or Scott's novels. But of all things I used to like to hear him read "Childe Harold;" and never shall I forget the way in which he read the description of the storm on Lake Leman. He took great pleasure in Byron, and Coleridge's "Hymn to Mont Blanc" delighted him. When anything touched his feelings as he read—and it happened not unfrequently—he would show it not only in his voice, but by tears in his eyes also. Nothing vexed him more than any kind of subterfuge or prevarication, or glossing over things.

'Once I told him of a professor, previously of high repute, who had been found abstracting some manuscript from a library. He instantly said, "What do you mean by abstracting? You should say stealing; use the right word, my dear."

'If he gave me my choice in anything, he could not bear indecision, and I had not only to decide, but to decide quickly. He thought that in trifles quickness of decision was important, and a bad decision was better than none.

'When my uncle left his study and came into the sitting-room, he would enter into all the nonsense that was going on as heartily as anyone, and as we sat round the fire he would often play some childish game, at which he was usually the best performer; or he would take a part in a charade, and I well recollect his being dressed up to act the villain, and very fierce he looked. Another time I recollect him as the learned pig.

'In times of grief or distress his sympathy was always quick, and no scientific occupation ever prevented him from sharing personally in all our sorrows and comforting us in every way in his power. Time, thoughts, purse, everything was freely given to those who had need of them.'

Some other reminiscences of his life about this time may also be here mentioned.

'He always required much sleep—usually eight hours.

'At one period he had to make so many commercial analyses of nitre for Mr. Brande, after preparing his lecture, that it was very late before he could turn to his own researches. More or less, he never was without some original investigation, and he would remain in the laboratory at work until near eleven at night, then he went to bed.

'He could spare little time for reading, except reading the journals and books of science, and the "Times" and "Athenæum." Then he always read the Scriptures more or less. When he was thoroughly tired and exhausted, which he often was, he turned to some story or novel that had a thread to it (as he said). This he found was a great rest. He did not take to biography or travels, but he would occasionally run through such books. They did not give him the complete relaxation he needed when thoroughly tired.

'He read aloud to his wife and niece, with great pleasure to himself and to them; sometimes he would read to them out of Shakespeare or Byron, and later out of Macaulay.

'It has been said that he liked to go to the theatre, and it has been concluded that he went very often; but really he went very seldom. He enjoyed a play most when he was tired, and when Mrs. Faraday could go with him. They walked to the theatre, and went to the pit, and it was the greatest rest to him. Sometimes, when she had a friend staying with her, he would go alone to some theatre, at half price. For many seasons he had a free admission to the Opera, and that he enjoyed very much; but he went only a very few times in the year, three or four. He was very fond of music, but he liked it to be good music. Before his marriage he played on the flute, and, probably to save expense, he copied out much music, which still exists; and he has said that in early life he knew a hundred songs by heart. After his marriage he had no time for the flute.'

If Faraday's scientific life had ended at this time, when he finished his higher education, it might well have been called a noble success. He had made two leading discoveries, the one on electro-magnetic motions, the other on the condensation of several gases into liquids. He had carried out two important and most laborious investigations on the alloys of steel and on the manufacture of optical glass. He had discovered two new chlorides of carbon; among the products of the decomposition of oil by heat he had found the bicarburet of hydrogen, or benzol; he had determined the combination of sulphuric acid and naphthaline, and the formation of a new body, sulpho-naphthalic acid; and he had made the first experiments on the diffusion of gases, a subject which has become, by the researches of Professor Graham, of the utmost importance.

According to the catalogue of scientific papers compiled by the Royal Society, he had had sixty important scientific papers printed, and nine of these were in the 'Philosophical Transactions.'

From assistant in the laboratory of the Royal Institution, he had become its director. He had constantly lectured in the great theatre, and he had probably saved the Institution by taking the most active part in the establishment of the Friday evening meetings.

But when we turn to the eight volumes of manuscripts of his 'Experimental Researches,' which he bequeathed to the Institution, we find that his great work was just going to begin. The first of these large folio volumes starts in 1831 with paragraph 1, and goes on, volume after volume, to paragraph 15,997, in 1859. The results of this work he has collected himself in four volumes octavo. The three volumes on electricity were published in 1839, in 1844, and in 1855 ; the last volume, on chemistry and physics, which contains also the most important of his earlier papers, was published in 1859.

Faraday's great work lasted for a quarter of a century. After the first ten years a break took place, caused by the strain that he put upon his brain. Giddiness and loss of memory stopped his work. These compelled the mind to rest comparatively speaking for nearly four years. After the rest was taken much more work was done. The pictures of these three periods will form the subjects of the three next chapters.

END OF THE FIRST VOLUME.